S0-BHX-984

Wonder Woman

Marketing Secrets for the Trillion-Dollar Customer

Iain Ellwood
with Sheila Shekar

First published 2008 by
PALGRAVE MACMILLAN
Houndmills, Basingstoke, Hampshire RG21 6XS and
175 Fifth Avenue, New York, N.Y. 10010
Companies and representatives throughout the world

PALGRAVE MACMILLAN is the global academic imprint of the Palgrave
Macmillan division of St. Martin's Press, LLC and of Palgrave Macmillan Ltd.
Macmillan ® is a registered trademark in the United States, United Kingdom
and other countries. Palgrave is a registered trademark in the European
Union and other countries.

ISBN-13: 978--0--230--20160--6
ISBN-10: 0--230--20160--1

This book is printed on paper suitable for recycling and made from fully
managed and sustained forest sources. Logging, pulping and manufacturing
processes are expected to conform to the environmental regulations of the
country of origin.

A catalogue record for this book is available from the British Library.

A catalog record for this book is available from the Library of Congress.

10   9   8   7   6   5   4   3   2   1
17 16 15 14 13 12 11 10 09 08

Printed and bound in Great Britain by
Cromwell Press Ltd, Trowbridge, Wiltshire

How we articulate the world is how we understand our own reality

Contents

List of Tables

List of Figures

About the authors

Iain Ellwood

Iain is Head of Consulting, Interbrand. He has many years' international experience, living and working in Japan, Hong Kong, the Netherlands and the United States as a marketing strategist for blue-chip companies. As a consultant he has led highly effective engagements for clients including Mitsubishi, BT, Orange, Barclays, British Airways, Nissan, InterContinental, RSA, Tesco, BP, Philips and UBS.

Before joining Interbrand, Iain worked at Prophet Management Consultancy. He led worldwide engagements creating high-impact solutions for customer proposition development, brand operationalization and marketing strategy.

Iain's focus is on driving higher margins and profits through putting the customer at the heart of the organization. This is achieved through effective and inspirational marketing strategy, brand operationalization and touchpoint development. Working with CEOs, his extensive expertise and knowledge have shaped corporate strategy and customer-focused propositions as well as internal brand campaigns that motivate and educate employees.

He is the author of *The Essential Brand Book* (Kogan Page 2001, translated into several languages) and a regular press commentator on marketing and branding issues for *The Economist*, the BBC, Sky News and numerous business magazines. Iain is a frequent international speaker on branding, innovation and communications. He also occasionally lectures for MBA courses at London Business School (LBS).

Iain is a member of the Chartered Institute of Marketing (MCIM); a member of the Marketing Society and a Fellow of the Royal Society of Arts (FRSA).

He holds a master's degree in social psychology from the University of London.

Sheila Shekar

Sheila is Senior Manager, Global Brand Management, Banana Republic. She is a senior marketing professional with over ten years of proven success developing consumer-insight-driven brand initiatives and marketing communication strategies for world-class brands, including Gap Inc., Banana Republic, UBS, Radisson Seven Seas Cruises and Visa.

Sheila currently leads international brand management efforts for Banana Republic, helping the brand expand successfully into a global lifestyle brand. Previously, she worked in corporate communications at Gap Inc.

Before Gap Inc., Sheila was a consultant at Prophet, a top global brand strategy consulting firm focused on Fortune 500 clients, with experience in developing and implementing brand-driven business strategies and marketing initiatives with global organizations across industries.

Prior to Prophet, Sheila created marketing communications campaigns and managed media and analyst relations for various global clients at Ketchum, a top-ten global PR firm.

Acknowledgments

I would personally like to thank the following marketers, business owners, colleagues, clients and friends for their insightful ideas and constant encouragement: Tom Agan, John Allert, Kate Ancketill, Gwynn Burr, Sue Carter, Ian Castello-Cortes, Rita Clifton, Charlie Colquhoun, Neil Duffy, Nick Durrant, Jez Frampton, Adrian Furnham, Rune Gustafson, Alastair Kingsland, Noel Penrose, Kate Rogers and Grant Usmar.

Thanks to my family – Jan, Andrew and Peter – for all their support. Finally, thanks to my father who saw the conception of this book but is no longer with us to see its fruition.

IAIN ELLWOOD

The author and publishers wish to thank the following for permission to reproduce copyright material: Andrea Learned for text from *Love Sweet Love: DHL Resonates with Women's Market* (2005); The Random House Group Ltd for a table from *The Female Brain* by Louann Brizendine, published by Bantam Press; Carat Media Agency for text from *Project Britain: Segmenting Simpletons* (2005); Verdict Research for text from Sean Hargrave in *Marketing Week*, 3 February 2006; The British Psychological Society for figures from D. J. Herrmann, M. Crawford and M. Holdsworth, "Gender-linked differences in everyday memory performance," *British Journal of Psychology*, 1992, **83**, 221–31, reproduced with permission from the Journal of Psychology, ©The British Psychological Society; Interbrand for data originating in 2006 (personal interview); Elsevier for text from the *Journal of Experimental Social Psychology*, 16, by G. Levinger, "Toward the Analysis of Class Relationships," 510–44 (1980), and for a figure and text from Byrne: *The Attraction Paradigm* (1971), copyright Elsevier; Springer Science & Business Media for a diagram from *Universal Principles of Design* by W. Lidwell, K. Holden and J. Butler, Rockport Publishers; SRI Consulting Business Intelligence (SRIC-BI) for a diagram from *MacroMonitor*, "Women as Financial Consumers: Gaining Ground," IV (2), January 1999; Palgrave Macmillan for text from G. Moss and A. Coleman, "Choices and Preferences: Experiments in Gender Differences," *Journal of Brand Management*, 9(2), November 2001: 89–98; The University of Chicago Press for text from Joan Meyers-Levy, "The Influence of Sex Roles on Judgment," *Journal of Consumer Research*, 14(4), March 1988: 522; nVision/Future Foundation for text from *Changing Lives: Media and Gender Survey, UK*; Associated Newspapers Ltd for material from Oliver Stallwood, "Why Women Find Parking So Tricky," *Metro*, 24 January 2005; Title Nine for material from its website titlenine.com, 2007 *Our History: US Sports Legislation*; Sage Publications for tabular material from J. E. Williams and D.L. Best, *Measuring Sex Stereotypes: A Thirty-Nation Study* (1982).

Every effort has been made to contact all the copyright-holders, but if any have been inadvertently omitted the publishers will be pleased to make the necessary arrangements at the earliest opportunity.

Introduction
Marketing to women makes business sense

Women friendly stocks outperformed the market by three times as much in the last couple of years and will continue to do so over the next ten years. (Goldman Sachs 2007)

Forget China, India and the internet: economic growth is driven by women. (*The Economist* 2006)

Women buy 80% of household purchases and 51% of online purchases. (*BusinessWeek* 2005)

By 2010, women are expected to control $12 trillion, or 60% of America's wealth. (*BusinessWeek* and Gallup 2004)

Marketing to women is a critical growth strategy for businesses and one that is entirely underleveraged. Or, as Tom Peters believes: "Women Rule!"

IT'S TIME TO CHANGE

This book encourages every business owner, employee and marketer to believe that marketing to women is currently the most effective business growth strategy. Women are the most financially attractive target audience and therefore marketing to them will accelerate higher shareholder value. Women buy the weekly grocery shopping, home insurance, books, household appliances, soft furnishings and linen, vacations, cars, furniture and more. CEOs and marketers first need to acknowledge that women really are the boss when it comes to buying. Second they need to understand that women are different from men in many psychological and behavioral ways that affect their relationships with brands. This book examines the key biological and brain differences between women and men and then defines more effective marketing approaches based on those insights. This is not about painting products pink or adding frilly patterns to the packaging; this is a new approach to marketing that challenges some of the sacred cows within the marketing discipline.

> Nike's executives have come up with strategies they hope will take advantage of the differences between how women and men conceive of sport, how they shop for clothing and shoes and what they think of celebrity athletes. (Wong 2001)

All of the tools and techniques in this book will help marketers to retune their brand experience specifically to women. The book is packed with case-studies and best-practice demonstrations of effective marketing to women. Women are the most important new business growth target and as Tom Peters (2005) confirms, "Women in my opinion are Economic Opportunity No. 1."

THE BUSINESS CASE FOR MARKETING TO WOMEN

It is important to begin by dispelling a few of the stereotypes to demonstrate the true business case for marketing to women.

Myth: Women are a niche market.

Truth: Women make over 80 percent of all purchases (in the US); and 50 percent of "male" purchases like cars, computers and so on *(BusinessWeek* 2004).

Myth: Men use online shopping more than women.

Truth: 62 percent of women, 38 percent of men shop online (US Census Bureau 2005).

Myth: Focusing on women will alienate men.

Truth: Successful marketing programs enhanced male customers' experiences and loyalty as well (Myers 2006).

Myth: Using feminine stereotypes and colors works for women.

Table I.1 Years by which women outlive men.				
	UK	*USA*	*China*	*World*
Number of women	30.6 million	151 million	700 million	3 billion
Number of years women live longer than men (average)	5.1	5.8	3.6	3.5

Truth: These often alienate women as much as men (Learned 2005).

Myth: Women only buy feminine products.

Truth: In the US, women buy 61 percent of major DIY products worth $70 billion (*Wall Street Journal* 2006).

Myth: Women aren't interested in sports.

Truth: Women bought more than 80 percent of NFL products in 2005, accounting for 40 percent of their total audience (NFL 2005).

3,262,585,132 WOMEN IN THE WORLD

There are over three billion women in the world today. These women also live on average 3.5 years longer than men till the average age of 66.47 years worldwide (see Table I.1).

There are many different segments of women but the three with the highest disposable incomes are Generation Y women, who were born after 1978; Generation X women, born between 1966 and 1977 and Baby Boomer women who were born between 1946 and 1965. Different markets have a different spread in the volume of these three segments. In the US, the spread is hourglass-shaped with more Generation Y and Baby Boomer women, while in the UK the segment size increases as the demographic increases, with slightly fewer Generation Y women and more Baby Boomer women. From a value perspective the Baby Boomers have a greater share of financial spend in both the UK and the US. This means that targeting the Baby Boomers in the UK is proportionally more financially attractive to businesses than targeting the other two segments, while the growth opportunities are clearly higher with the Generation Y segments (see Figure I.1).

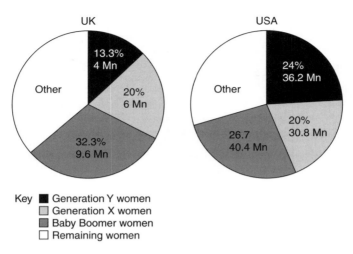

Figure I.1 Most financially attractive female customer segments (Sources: CIA 2006; Learned 2004; Carat 2005).

As we have just seen, the financial evidence for marketing to women is clear and there are many other examples throughout this book. Understanding the female mind is therefore paramount to increasing sales and profitability for all businesses. This is not a radical shift that is part of some new-millennial feminization; it has always been the case. Women have looked after domestic arrangements since prehistoric times. It is easy to list their scale of influence from healthcare and insurance through household goods and foods to vacations and clothing. Women purchase more than 75 percent of the over-the-counter drugs in the US (Gordon 2002). In fact it is easier to list the male-dominated purchases characterized as the three T's: technology, tools and toys (boys-toys, that is – like televisions and hi-fi). In most other areas women are the primary purchasers yet the focus of marketing has historically been on men. Car purchases also have become more female-dominated. In the US, the percentage of women buying new cars has risen from 30 percent to 50 percent in the last fifteen years, while in the UK a study undertaken by Condé Nast highlighted that women now buy 60 percent of cars. Similarly, Wyndham Hotels increased its share of women business travelers by 59 percent when it instigated its "Women On Their Way" service program (Myers 2006). All these examples demonstrate the revenue growth opportunities of marketing to women. This is confirmed by a recent Goldman Sachs study of women-friendly businesses that highlighted that they generated three times the return of the wider stock market.

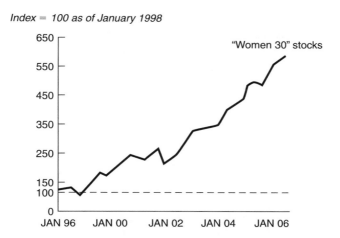

Figure I.2　Performance of "Women 30" female-friendly global stocks (Source: Goldman Sachs 2007).

Kevin Daly at Goldman's tracked a "Women 30" basket of global stocks that have tailored their products more to women over the past eight years. He emphasizes that "Women's income is rising relatively to men's and he expects that the out performance of female friendly stocks will continue over the next ten years" (see Figure I.2; Goldman Sachs 2007).

This clear financial case for marketing to women is increased further by the scale of this untapped market waiting to be exploited by progressive businesses. These women account for a large proportion of the total population and because they have a significantly higher propensity to buy than men or other segments they represent the perfect opportunity for any business and marketer. These women have increased their earning power dramatically and marry later in life or stay single. Over the past fifteen years men have reduced their influence on household finance while women now manage more household's finances than men. Oppenheimer Funds research (2006) into "Women and Investing" confirms that "In dual income families, 30% of working women out earn their husbands." More and more women are working alongside men in traditionally defined male jobs such as engineering, finance and management. Women earn more 50 percent of all accounting degrees and almost half of all law degrees (Krotz 2006). In the US in 2005, women make up 48 percent of the working population, up from 34.2 percent in 1983 (US Census Bureau 2003). There is a continuing increase in the number of women in the boardroom, although clearly the

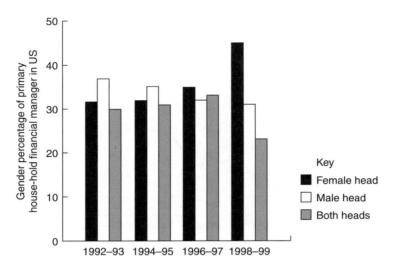

Figure I.3   Primary household financial managers in the USA by gender (Source: MicroMonitor 1999).

starting point was a low base. Last year one new appointment in five to a FTSE 100 company was a woman (Buckley 2004), compared with one in ten only two years ago. There are however still only two FTSE 100 CEOs, Marjorie Scardino at Pearson and Jose Maria Bravo at Burberry. Women can now be found at the pinnacle of every profession and this is no longer just a Western phenomenon. US Secretary of State Condoleezza Rice may have topped *Forbes Magazine*'s 2006 Top 100 Women and Xerox's CEO, Anne Mulcahy, is at No. 6, but China's Wu Yi, the health minister, is at No. 2, and Xie Qihua, the chairman of Shanghai Baosteel, is at No. 14. A MacroMonitor study of the US has shown that the primary household financial manager is now the woman in the house and not the man. These women hold the purse strings of the entire household income and spend accordingly. They are therefore in financial control of the majority of outgoings and are hence the primary brand decision-maker (see Figure I.3).

GENDER-BASED MARKETING STRATEGY

This book explores the differences between women and men and defines better ways to engage with, market to and satisfy women customers. The emphasis is on the essence of women but this does not mean all women are alike and all men the opposite. It is more insightful to describe the essential characteristics to illustrate their impact on marketing to women. The reality is that there is a spectrum of feminine and masculine attitudes and behaviors that covers all regions and markets. This book is about gendered attitudes and buying behaviors.

Any book like this is bound to be controversial for some people, either women or men. They may see the differences as unrealistic or even offensive. But the quantitative research evidence has proved across a range of subjects from genetics, neuropsychology, physiology and sociology that there are substantive differences between women and men that impact on their attitudes and behavior towards branding and purchases. Equally there are still a lot of myths and stereotypes about the differences between women and men. Some of these are a product of our current social culture and may well change. They have been changing dramatically over the last forty years and will continue to do so equally rapidly.

"Gender" is a term used to describe a person as female or male. It refers to the socially constructed difference between women and men – "sex" being the biological difference. This book is not about the politics of gender; while the latter is important, it is too theoretical for a practical guide for marketers who want to improve their business. This book is unashamedly driven from the point of view that self-identity is a blend of genetic formation and social construction. It acknowledges that people have a degree of self-determination over their representation. It is not a discourse on the state of modern feminism but is based on observable feminine behavior and its impact on marketing activities. Given the foundation of a patriarchal society, gender in culture is often seen as a deviation from a genderless universal norm. In the case of the female gender we can find evidence of this through specific forms of culture:

- *Woman's Hour* on the radio
- the women's page in the newspaper
- women's sessions at the gym

These products and services are designed specifically for women. But this book is about purchases that are non-gendered. It describes products that are suitable for both genders and not just women-specific products like feminine hygiene, or cosmetics brands. It is easy to see how popular culture overemphasizes feminine or masculine traits. Television programs illustrate popularly ascribed attitudes and clearly highlight several of the underlying

Table I.2 Typical feminine and masculine characteristics.

Feminine	Masculine
Dialogue and conversation	Action
Family	Professional network
Relationships	Achievement
Feelings and emotions	Activities and thoughts
Personal	Public
Home	Work
Community	Individual
Equality	Hierarchy
Soap opera	Current affairs
Romantic comedy	Action film
Novel	Historical book
Woman's Hour	Sporting news

essential characteristics of women and men. Just because these may seem obvious does not make them invalid. However, this book uses scientific research to provide a quantitative foundation that supports the primary and secondary characteristics of women. There is also evidence that dismisses female characteristics that are simply urban myths. Table I.2 lists typical female and male characteristics that will be re-examined throughout this book.

There are of course many similarities between women and men as well as differences. The reason that it's important to focus on the differences is because those differences are accentuated during the marketing and sales processes. Their very nature has an amplifying effect on specifically feminine and masculine traits and behaviors. This is not a solution to a "better or worse" discussion about women and men, but recognizes that we are all just different at different things. However, in marketing the focus has for too long been on a masculine version of the world that has underweighted the importance of women as consumers. There is a large body of commercial and academic research into the representations of women in marketing, primarily in their definition within advertisements. Most of this research has concluded that women are more frequently objectified within advertising than men. The resulting unhappiness this brings women means that that these marketing campaigns are actively alienating these women on a daily basis. This marketing attitude will need to change as businesses recognize the fundamental truth articulated in this book. Women are decision-makers in more than 80 percent of purchases in the Western world and anyone who wants to increase business performance would be a fool to ignore this fact.

MARKETING TO WOMEN

Marketing to women is not a zero-sum game for a business, in which marketing to women means that fewer men buy things. It is simply a rebalancing of the marketing landscape. Nor is it a binary debate where women are either girly or gay. Marketing to women covers a spectrum of attitudes and behaviors that demonstrate an increasing emphasis on feminine traits over masculine ones. A quick look at women's primary characteristics illustrates how powerful these arguments are:

- women are more perceptive and have higher empathy for others
- women are more interested in creating long-lasting relationships
- women are better communicators than men
- women are more engaged with managing their self-esteem
- women invest more energy into nurturing and sharing their emotions
- women are better able to analyze complex messages and build the bigger picture

For anyone working in marketing the list provides unequivocal direction that will result in superior performance marketing. This is neatly characterized by Faith Popcorn and Lys Marigold (2001) as: "Women don't buy brands they join them." Addressing the authentic reality beyond female stereotypes is imperative if businesses are to maximize their financial gains. This means understanding the wider context of female consumption habits that have been succinctly summarized by Wendy Gordon (2004) as follows:

Marketing to women is also a shift from male to female values:

- from competition to collaboration
- from rational thought towards emotional and intuitive
- from focused activity to understanding the wider context

Table I.3   Key differences produced by women's and men's brain hard wiring.

Women	Men
Affectionate	Adventurous
Multitasking	Lazy
Gentle	Aggressive
Vague	Clear-thinking
Sentimental	Coarse
Dependent	Courageous
Emotional	Unemotional
Submissive	Dominant
Selfless	Egotistical

In order to appreciate these behavioral differences it is important to comprehend the psychological foundations of women and men's brain hardwiring. Williams and Best have highlighted the key differences in the academic bible, *Psychology, Brain, Behavior and Culture* (Westen 2002; see also Table I.3).

One of the challenges with marketing to women is for marketers to define what is genuine and what is just a stereotype or convention. There are many examples of overtly feminine marketing techniques that recourse to pink fridges and flowers on toasters. They emphasize both the traditional feminine domestic role and the stereotypical feminine social leitmotifs. In order to abstract this into something more contemporary we need to identify authentic psychological differences rather than those that are simply represented by the media. The clever corporations have now acknowledged that if they are to gain more from their customers they must develop a stronger relationship with them: "In the knowledge based economy, workers will be valued for their ability to create, judge, imagine and build relationships" (Schwartz 2006). Women prefer long-term relationships with their brands more than do men. Women have historically played a lesser role in product and marketing development. The dominant force in marketing departments has been masculine – testosterone-fueled and achievement-driven. These attributes are, as we shall see, just the things that women find not particularly motivating or influential to their purchasing decisions. The challenge

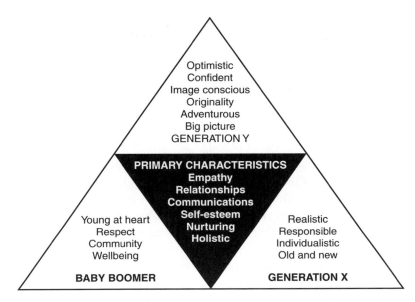

Figure I.4 Primary characteristics for all women and the additional secondary characteristics for Generation Y women, Generation X women and Baby Boomer women.

for marketers is to transform their working practices and enjoy the financial benefits of marketing to women.

HOW TO GET THE BEST FROM THIS BOOK

The first part of this book, Chapters 1 to 3, provides the statistical scientific evidence for the differences between women and men and their consumption behavior. This gives marketers a deeper understanding of the genuine differences between women and men. The middle part of the book, Chapters 4 to 6, is a groundbreaking analysis of specific female segments and the marketing strategies and tactics required to satisfy these segments. This looks at the primary characteristics for all women and the additional secondary characteristics for Generation Y women, Generation X women and Baby Boomer women, the three most influential and affluent female segments of the population (see Figure I.4).

The final three chapters provide specific marketing tools and techniques designed to deliver higher-performance marketing to women. These tools are straightforward, proven and can be used immediately by any marketer, and are illustrated with numerous case-studies.

Chapter 1, "The female brain," provides the scientific evidence for the differences between women and men. There are some obvious biological differences that distinguish women from men. But there are far greater unseen neurological, psychological and hormonal differences that create most of the significant differences in women's consumption attitudes and behaviors. Women are better at putting themselves in other people's shoes than are egocentric males. This means that they are better able to actively listen, empathize and respond to customers, which results in stronger and more long-lasting customer relationships. These lead to a preference for emotional decision-making and communication methods that rely on previous "brand memories" and the role that our memories play in categorizing and choosing those brands.

Chapter 2, "Women's primary characteristics," defines the most effective ways to understand what women want, and want to feel about your brand. It explores the different kinds of emotions that all women experience, and establishes the six foundational characteristics of women. It looks at the uniquely female functional and emotional needs and goals. It includes a large number of case-studies that illustrate best practice in marketing to those core female needs.

Chapter 3, "Researching women's needs," defines the most effective ways to research and target specific demographic, attitudinal and life stages of women. Without this refined level of targeting it is likely that any marketing activity towards women will be too generic and superficial. Marketers need to gain deeper qualitative insights into what drives female purchase behaviors. The chapter concludes with a description of the latest research techniques that will help uncover latent needs and desires of women. These include psychographic profiling, ethnographic observation techniques and semiotic analysis.

Chapter 4, "Generation Y women," explores the specific secondary characteristics and behaviors of this segment of women. This is the youngest of the three financially most attractive female segments. They are optimistic and believe life is for living. They are confident and have high self-esteem. These women are able to move fluidly across traditional boundaries like the work–life balance or the ability to buy both premium and value brands in combination for example. They no longer conform to a traditional and dogmatic purchasing regime. They are in control of their lives and play the game of life to their advantage. It illustrates how brands use strong marketing and visual imagery to attract this specific group of women.

Chapter 5, "Generation X women," explores the secondary characteristics and purchase behaviors of women born between 1967 and 1977. These women are independent, more realistic and conventional in their attitudes. They bridge a transition point in society and have to try and balance traditional and modern lifestyles. They have a strong sense of responsibility. They are often described as part of the Lost Generation. These women are more serious than Generation Y women, and marketers need to emphasize reality to these women rather than utopian futures.

Chapter 6, "Baby Boomer women," focuses on the additional secondary characteristics of women born between 1946 and 1966. These are currently the wealthiest group of people on the planet. They have the time, energy and money to buy delightful products and services. Baby Boomer women are young at heart. They grew up in a time of growth and optimism. They have few debts or regrets and want to continue living their lives to the full. They are fit, active, and are eager to avoid the decline they witnessed in their parents.

Chapter 7, "Marketing communications," identifies the key differences required to communicate effectively with women audiences. It emphasizes the differences in women's verbal, visual and cerebral skills to illustrate more effective marketing messages and communication channels. These include more sophisticated use of vocabulary, tone of voice and analogies to enrich communications. This acknowledges the higher level of language and communications that women are capable of understanding and using. It defines the best practice for design and other visual elements when communicating with women. This chapter provides examples on how to develop more effective and persuasive marketing themes and messages; it concludes with specific changes to major marketing communications channels to make them more effective to women.

Chapter 8, "Effective brand experience design," defines the best way to make women feel good about your brand. Women are significantly more interested in how they "feel" about a brand than how they think about it, which is a much more masculine way of choosing brands. It describes the feelgood framework and identifies the key tools that can be used to create enhanced brand experiences that are more appropriate for women customers. Making women feel good is the key to them loving your brand. Women are cognitive of a more holistic experience of the world than men and this means that brands need to have a multidimensional approach to satisfying their needs and making them feel good.

Chapter 9, "Touchpoint improvement," identifies the high-impact touchpoints for successful marketing to women. It establishes that the most effective marketing strategy is the ability to build relationships with women through an integrated touchpoint experience. Their inherent expertise at relationship-building and need for socialization is a marketing dream – if brands are able to commit to the relationship. This chapter describes an effective six-step touchpoint improvement process that can build more powerful relationships with women. It identifies key characteristics and techniques that need to be deployed at different stages of the customer relationship cycle. The chapter concludes with an overview of developing the business return on investment (ROI) case with regard to developing touchpoints specifically aligned with women's attitudinal and behavioral preferences.

PROVOKING NEW THINKING

The scientific context of this book provides the intellectual foundation for new strategies, tools and tactics to create more effective customer relationships with women. There is an empirical business case for marketing to women as a highly effective growth tool. Marketers need to be vigilant against myths and stereotypes of women customers; they require more insightful research and marketing approaches in order to be effective. Female consumption patterns are hallmarked by holistic marketing and collaborative dialogues with them. The intellectual and emotional content of brands is prioritized by women over the functional and rational. It is therefore worth reading all of the brain and biological differences chapter prior to beginning the chapters on specific segments of women and the subsequent marketing strategies that are more effective. However, if the reader is already conversant with the research that underpins this then the later chapters can be used as a toolbox that can be accessed in an ad hoc manner. The strategies, tactics and tools have all proven highly successful through use with major corporations around the world and across all sectors. They are leading-edge however, and some may challenge current marketing practice. They are supported by case-studies and scientific evidence to ensure that the marketer can maximize their immediate impact.

The book draws on decades of experience of working with the world's leading brands to help make them more appealing and differentiating to women customers. Many of the examples are drawn from personal experience of working in the US, Europe and Asia. As a topic, marketing to women has fascinated us on many levels. As a customer group they are powerful but they are targeted less well than men. This is partly because marketing teams are often split by gender, with the majority of senior marketers being men. As a customer segment to research, there has been little commercial tailoring of research techniques – despite the obvious differences in their attitudes and behaviors. Finally, as an audience for marketing strategists, women with their higher relationship-building and communications talents represent the guiding beacon for the future of the industry.

We hope that the book is provocative and that it prompts you to think differently about marketing to both women and men in the future.

1

The female brain

This chapter provides an overview of the scientific evidence of the differences between women and men and how they affect their brand perceptions and purchase behaviors. We start from a child-developmental standpoint and identify the hybrid of the nature and nurture opinions to support a modern perspective of gender-based marketing. This framework allows us to identify the issues that provide deep insights into the differences and similarities between women and men. The first of these issues is the differences in the brain and central nervous systems that result in the different hard wirings of the female and male brains. These govern how we gain our brand perceptions. Second, it describes the effect of hormonal differences, especially during childhood development, on female behavior. Next, we identify how the mind links perceptions with the creation of powerful brand memories: how they are interpreted, organized and retrieved. The chapter goes on to link these to the emotional purchase behaviors of women toward brands that characterize many of the essential differences between women and men. The evidence is that women's behavior is driven more by emotional factors than men's and that their behavior is likely to vary more widely than men's. To summarize, the key differences in behavior between women and men are driven by their DNA, the hard wiring of their brains, the effects of hormonal differences and their upbringing. This chapter sets out the fundamental differences between women and men and why they matter, and shows how they might influence how we market towards women.

OBJECTIVES

- Identify the biological, neurological and psychological differences between women and men

- Define the process by which women create their own brand perceptions and memories that lead to higher brand appeal
- Identify the role that emotions play in a woman's purchase behavior that leads to stronger brand loyalty

WOMEN'S BRAINS ARE DIFFERENT

Women and men share 99 percent of the genetic coding of their 30,000 genes. But that one percent has a dramatic effect on the differences between the two genders. It influences every single cell in our bodies from bones and muscles to psychological feelings and emotional responses and finally even to the length of our lives. Much of our knowledge about brain differences comes from observing the different brainwave patterns identified using magnetic resonance imaging (MRI) scans. These have been especially valuable in observing differences between women's and men's brains while undertaking identical tasks. For example, a woman's brain is on average 9 percent smaller than a man's, BUT this does not equate to any difference in intelligence (Brizendine 2007). MRI scans indicate that women have more tightly packed brains, with 11 percent more neurons than men's, especially in the cerebral cortex regions that influence language and hearing. The stereotype that men think about sex far more often than women may be based on the fact that the area of the brain that manages sexual drive is twice the size in men than in women and the areas of action and attack are also much smaller in women (Brizendine 2007). This may be one reason why women can hold platonic relationships with the opposite sex much better than men, who tend to invest most relationships with a sexual dimension.

Low laterality; connecting the two hemispheres

It is well known that the brain is made up of two hemispheres with distinctive characteristics that fundamentally shape our brand perceptions and purchase behavior. A woman's brain is much better connected across the hemispheres, which means it has a superior cross-function processing ability to that of the humble male. The left cortex area in particular manages language competency, while the right cortex manages the tonal/musical quality of speech. The connection between the two hemispheres of the brain has a strong effect on the differences between women's and men's

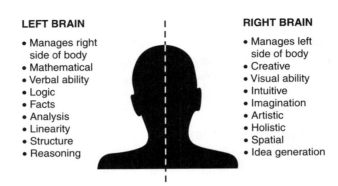

LEFT BRAIN

- Manages right side of body
- Mathematical
- Verbal ability
- Logic
- Facts
- Analysis
- Linearity
- Structure
- Reasoning

RIGHT BRAIN

- Manages left side of body
- Creative
- Visual ability
- Intuitive
- Imagination
- Artistic
- Holistic
- Spatial
- Idea generation

Figure 1.1 Typical left and right brain characteristics.

purchase behaviors. Typical right and left brain characteristics are shown in Figure 1.1.

The level of connection between the two hemispheres of the brain is known as laterality. A brain with low laterality communicates more between the two halves than one with high laterality where the activities are more discreet. Women tend to have brains with lower laterality and with better connections across the hemispheres. This is often evidenced during MRI scans, or when one side of the brain is damaged; women are less likely to suffer complete loss of a function than a man. The corpus callosum, one of the connecting tissues in the brain, is generally larger in women than men, thus providing more rapid inter-hemispheric transfer of information. Women have a better integration from all their senses and thoughts because of this greater crossover and sharing between the two hemispheres of the brain. In other words, women are "better at getting it all together" or getting the bigger picture and a wider understanding of consequences of their actions. The hard wiring of the female brain also favors emotional sophistication: "The connections across the emotional systems in a women's brain are stronger than the rational ones" (LeDour 2005). Higher exposure to male hormones (androgens) during fetal development is said to produce a greater specialization between the two hemispheres in the male brain. The right hemisphere is dominant in men and controls spatial and visual abilities. Conversely, women's brains have less specialization of the brain hemispheres but still have the left-hemisphere dominance that controls verbal abilities. Moir and Jessel (1989) have argued that this specialization accounts for several stereotypical behavioral traits; thus for example because men have a higher separation between the left (emotional) and right (verbal) areas of the brain they are less able to talk about their emotions.

> Physiological differences in the structure, make-up and connections across the brain cause women and men to process perceptions and behaviors differently. This means that women make sense of a brand in ways significantly different from those employed by men.

Because women are left-hemisphere-dominated, their higher verbal skills often mean they are interested in careers in journalism, writing and the study of languages, while right-hemisphere-dominated male brains, with a superior spatial ability, often direct their owners towards architectural or engineering subjects. However, in the female brain the right hemisphere specializes in emotional information – a specialism that the male brain lacks, having its own specialism for spatial reasoning. This confirms the stereotype that women are better at emotional issues while men are better at spatial ones. A study by Turhan Canli (2002) confirms that "women's brains are wired better both for feeling and recalling emotions than [those of] men."

Marketing to girls

School exam results have consistently shown that girls are better than boys earlier in their school careers. This early advantage of girls tends to affect boys and causes them to be less confident as they progress through school. Girls are generally better at languages, the arts and English (subjects that don't require great spatial skills). They are particularly good at learning foreign languages. The boys are better at mathematics and physics, which don't require great linguistic skills. It is only as boys begin to study mathematics and science that they can regain their equality or any sense of superiority. Because the girls are more articulate, they are able to maintain some advantage even through the difficult teenage years. As surly boys grunt their way through these years, the girls are chatting about who said what to whom during class. Marketing communications to girls can therefore use more sophisticated language, while boys will be better with punchy headline statements only. Typical adverts for Barbie and Action Man dolls demonstrate this clearly:

"Elina Barbie – the FAIRYTOPIA™ adventure continues; join the magical fairy fun"
and
"Action Man – big enough to handle any mission with gripping hands"

The key differences in the female brain have resulted in significant differences in abilities and brand perceptions that lead to different purchase behaviors:

Advantages of the female brain

- Verbal and nonverbal communication
- Reading your partner or listener's feelings
- Higher emotional intelligence
- Mothering and nurturing
- Making friends
- Being better at sorting and matching type tasks on objects of the same color
- Being better at performing precision manual tasks
- Being better at recalling landmarks from a route
- Greater self-control. Women are better at self-control; this is a fundamental ability to building empathy. The ability to switch off your own feelings, and become less self-centered is a prerequisite for empathizing. If we remain egocentric it will be almost impossible for us to develop significant empathy with others
- Women are better able to focus on the big picture than men

Advantages of the male brain

- Making and using tools
- Trading
- Power and competitive attitudes
- Social dominance
- Tolerating solitude
- Aggression
- Leadership
- Focus on detail

These major differences need to directly influence marketing, design and communications techniques towards women. Women are better able to grasp the big-picture emotional benefit and link that to the detailed product features. Men rarely get beyond the headline grabbing statement, unless it's a power statistic. The language used towards women should focus on the emotional feeling that the product will deliver, rather than the technical performance of the product. Women prefer brands that enable or enhance socializing and group bonding. They choose brands that stress inclusiveness, while men prefer brands that are emblematic of supremacy and status. They favor brands that characterize singularity. These will be defined in much greater detail later in the book.

Femininity is a national characteristic

The Dutch anthropologist Geert Hofsteede studied the masculinity or femininity of nations based on a survey encompassing over 53 countries from the Old World and the New World, rich countries and developing ones. His detailed survey and interviews with employees and customers enabled him to build up a picture of attitudes towards general subjects like work, family life, hobbies and entertainment that he later analyzed for any feminine or masculine bias. He found a statistical difference between how masculine a country was (including both men and women) and how feminine it was (in terms of demonstrating the typical gender traits). His key finding was that the more strongly masculine the country was, the greater the difference in masculinity between men and women. A country like Sweden with a low masculinity score demonstrated highly feminine tenderness and nurturing traits and showed little difference between men and women. In contrast, in a country like Japan that demonstrated a strong masculine index there was a much larger difference between men and women's behaviors. Marketers need to identify which countries require significantly different marketing programs for men and women as well as the ones that can use a more inclusive approach (see Figure 1.2).

This insight on different markets is crucial to balancing the global–local appeal of any brand. In order to balance the global marketing efficiencies with local appeal and responsiveness you need to define which parts of the brand are non-negotiable and which parts can be dialed up or down within individual markets. It is crucial to avoid the flexible approach that allows countries to pick and choose which parts of the brand positioning they want to use. You must also shun a politically acceptable solution of blandness

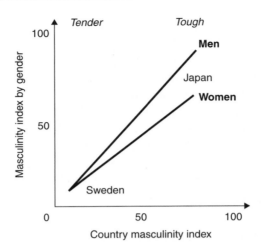

Figure 1.2   Masculinity indices for Sweden and Japan by gender.

that fails to deliver a crisp and polarizing brand positioning that will confer commercial benefit on the business. In our experience, the dial-up, dial-down model that requires all parts of the brand positioning to be used in all markets but with a different emphasis provides the optimal balance for global–local brand marketing.

INFLUENCING BRAND PERCEPTIONS AND PURCHASE BEHAVIORS

Emotional purchase behaviors are a direct response to perceptions of external stimuli that our senses receive such as sight, sound or touch. They involve a physiological receptor like an eye reporting to the brain, which is then evaluated against our memory of previous experiences, and which in turn drives an emotional and external behavioral response. There are several theories (notably James–Lange and Cannon–Bard) on the order that these events take under different circumstances, but the contemporary view is that the definition by Cannon and Bard is the most accurate. Their research asserts that external stimuli, for example a car driven recklessly towards you on a street, or a highly seductive impulse-buy checkout advertisement, produce both an emotional experience and a behavioral response in parallel – rather than sequentially as James and Lange have asserted. Our sensory organs are the conduit for these physiological emotion-inducing

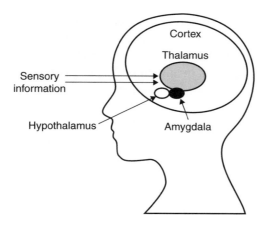

Figure 1.3 Hypothalamus, limbic system and cortex regions of the brain.

stimuli. Our sight, hearing, touch, smell and taste receptors are the starting point for any externally initiated emotional response. For brands, this means that the brand experience or communication around that experience is the battleground for generating stronger emotional responses than the competition's brand. Clever advertising, as we have all experienced, generates powerful emotional responses that can be turned into significant purchase behaviors.

Neuropsychology of emotional brand perceptions

The key to influencing our emotional brand perceptions is through harnessing our central nervous system. There are three principal parts that interact to produce and regulate our emotional perceptions and behaviors: the hypothalamus, the limbic system and the cortex (see Figure 1.3).

Limbic system

The part of the limbic system called the amygdala plays a key role in interpreting sensorial stimuli as emotional buying behaviors. It receives information from the sensory organs like the eyes via the hypothalamus and the wider cortex areas of the brain. The amygdala is often thought of as the arbiter or definer of our emotions. It is also heavily involved in the interpretation of other people's emotions through a similar process of assessing other people's emotions via body language such as facial expressions. A woman's amygdala is more easily triggered by emotional stimuli than a man's and this makes women's memories stronger because they have a richer emotional content than men's. This is great news for you because this means that more emotional marketing campaigns are more likely to be positively engaged by women, who will have a stronger memory of them than men (Brizendine 2007). The key to activating the female brain is to provide rich experiences that can then be stored as a multicolored, three-dimensional emotional memory. Men on the other hand struggle to interpret and incorporate so much emotional content and therefore have weaker, more monodimensional memories of a brand experience. Men do remember some emotional experiences as well as women, but they are more likely to be connected with anger, aggression or achievement. These classic male personality attributes can provide vivid memories because they are so directly linked to core masculine behavioral traits. This means that marketers can increase the effectiveness of their marketing campaigns by activating female emotions through expressing more feminine traits. Women will also have much stronger brand memories of a brand that correlate directly with their own feminine attitudes. Again, this means it is crucial for marketers to gain a deep understanding of target audiences underlying emotional traits.

Women are much better at reading nonverbal cues such as body language and their emotional response has a higher level of sophistication. They are also better at sensing other people's emotions and adjusting theirs to respond positively to this situation. This is because they have higher empathy skills and are more able to put themselves in other people's shoes. Their higher drive for collaboration means that they also respond better by subverting some of their desires in order to create a positive resolution. Given these different ways that women receive and respond to emotional stimuli as compared with men, it requires more sophisticated research and analysis to define the true nature of women's desires and needs.

Without the amygdala, we would not be able to understand the emotional significance of stimuli we receive. In cases where this part of the brain has been damaged, patients have been found to be unable to recognize previous, highly emotional stimuli, such as a loved one's face or the inherent danger that a gun has in our minds. They were unable to navigate the world on an emotional level and treated all things as emotionally equal.

Two distinctive brain processes occur when an emotional stimulus is received. There is first a primitive and immediate process, second a more sophisticated and measured one. It can be argued that women use the second, more considered process more often than men, who tend to be less considered and make more snap decisions. The immediate response involves the thalamus, which processes the sensory information directly to the amygdala and this in turn generates an immediate and somewhat reactive response. For example, a sudden bright light will make us cover our eyes, or encountering a dangerous bear in the woods will cause a fight-or-flight response. These are basic responses designed to avert immediate physical danger without deliberation or "thinking" about them.

The second, more considered response is also driven by the amygdala. When it sends its initial response to the thalamus for immediate action it simultaneously sends the information to the wider cortex area. This area of the brain examines the stimuli in greater detail, compares it with previous experiences and evaluates it against potential responses to provide the most effective one. Once these deliberations (taking only nanoseconds) have occurred the cortex sends a response to the amygdala to provide an external response to the stimuli. The immediate physical reaction may be to remove your hand from the hotplate (based on a primitive physiological response), the secondary reaction may be to blow on it or put it under running cold water (drawing on our memories). In other words there are two ways that brands can generate emotional responses to their experience: (1) the immediate and physiological and (2) the more considered that involves reference to our previous experiences and memories. As women are more sophisticated at the secondary type of response, effective marketing needs to link closely to previously experienced situations and memories.

> Emotional rather than functional brand perceptions are more strongly processed into behaviors by women's brains than men's.

Obviously the basic response is quicker (shorter neurological pathways) but reacts to larger and blunter stimuli such as large-scale danger or pleasure. The second response, with its greater level of processing and interpretation, is slower and reacts to a finer, more granular level of stimuli. For businesses, the greater and more immediate effect is desirable but difficult to achieve. Disruption campaigns and shock-tactic communications can be more effective with men because their brain uses more of the primary response mechanism than does women's.

Modern consumers are well experienced at interpreting advertisements and brand experiences and they are unlikely to project the substantive stimuli that will yield a primary emotional response. Some advertisements use shock tactics to try and achieve this result. Anti-smoking, drinking or drugs campaigns often use gruesome images and stories of sufferers to create these basic responses. Benetton, the clothing company, used powerful images by photographer Mario Testino as a shock tactic. He combined often highly contradictory images of life and death to elicit extreme primary emotional responses.

There is no evidence that the primary or secondary mechanisms produce a longer-lasting memory, although logically the more substantial stimuli produce a quicker but more diffuse response, while the accuracy and specificity of the more considered responses should prove more memorable and therefore significant for customers. The added specificity is also more likely to help a brand differentiate itself emotionally, rather than relying on category or industry-defining characteristics. This argument therefore suggests that brands should concentrate on the secondary type of emotional response, particularly in mature categories or markets. However, if the aim is simply to stimulate the entire market with a new product or service then focusing on market fears and pleasures may be a more effective route for brand experiences and advertising. The way these responses are enacted is through the hypothalamus.

Role of the cortex

The cortex plays several roles in managing our emotions. As we have seen, one role is to assess incoming stimuli and determine the level and type of response required. A second role is a reflective one, helping us to interpret physical activities in our body; it interprets our rising blood pressure, physical agitation and increase in heat as feeling angry. The third role the cortex plays is most crucial in defining and interpreting our and others' emotions; it is responsible for defining and managing the facial expressions of our emotions. The right and left hemispheres are used differently, with the right hemisphere playing more of a role in interpreting the emotions of others. This is a crucial capability that allows women to empathize with others rather more than do men with their more egocentric attitudes. Research has shown that people who are more left-hemisphere-dominated are more likely to experience positive emotions, while those with greater right-hemisphere activity are more likely to experience more

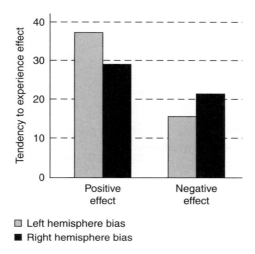

Figure 1.4 Left and right hemisphere biases towards positive and negative emotions.

negative emotions (see Figure 1.4; Tomarken *et al.*, quoted in Westen 2002). Women appear to experience emotions more intensely than men. They are also able to better read people's emotions more accurately. As this is a key skill in interpreting brand experiences, it is likely that women are more sophisticated at interpreting and memorizing brands.

Classifying behavioral triggers

Our emotions can be divided into a number of key categories, and marketers need to define which emotional response they are trying to achieve with their female audience. Without clarity on the target emotional response, it is likely that the brand experience or advertisement will be confusing, diluted or ineffective. The first basic division of types of emotions is positive and negative, and both of these can be equally powerful and intense (see Figure 1.5). Positive emotions are processed more by the left frontal lobes and drive pleasure-seeking and gregarious behavior in people, while negative emotions are processed more by the right frontal lobes and drive avoidance behaviors. This simple division can have an enormous effect on the way people organize their lives. Those with a higher disposition to positive emotions are more likely to be pleasure-seekers, to take risks and to experiment with new and exciting things – like Generation Y women.

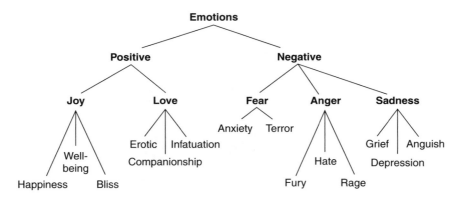

Figure 1.5 Taxonomy of emotional levels.

Those with a higher disposition to negative emotions are more likely to be risk-avoiders, less social and more cautious – like Generation X women. Part of our disposition is inherited, with emotionally negative parents more likely to pass on this trait than those that are emotionally positive (Watson and Tellegen 1985).

The next level of basic emotions can be divided into five core emotions. Marketers need to identify which of these is likely to underpin the externalized need for their target audience:

- love
- joy
- anger
- sadness
- fear

While different cultures may interpret these differently, the same concepts will still exist. Beyond these five, there are a range of emotions that may be interpreted differently depending on the context of culture. However, for modern, Westernized society they are sufficiently similar for most brands.

Synchronicity

Synchronizing the positive associations of your brand with attitudinal nodes already existing in a woman's memory will maximize appeal for her. Defining a specific brand-positioning requires clarifying the woman's mind map

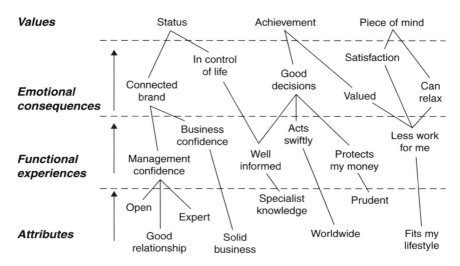

Figure 1.6 Laddering used to identify higher-order brand attributes.

of all relevant brand attitudes. Rather like a stereotypical therapy session, this involves using appropriate research techniques to unpack the mind and a woman's thoughts in relation to a brand. By repeatedly probing words, visual imagery and brand attributes we can begin to paint an accurate picture of the target woman's mind map. Laddering techniques can also be used to prioritize those brand attributes, from relatively insignificant product features to higher-order concerns, values and beliefs (see Figure 1.6).

Women have a much higher empathy quotient than men, which means that stronger emotional content can be used when marketing to women. The second benefit of having a high empathy quotient is that they are able to adapt better to changing situations and emotional tensions. This is because women can subvert their feelings more quickly and fully than men are able to. This does not mean they do not have these feelings of anger or delight, but simply that they are better able to hide them in order to ensure that their response does not further antagonize the situation. You need to demonstrate strong levels of empathy in communicating with women in order for them to believe any brand relationship is going to be worthwhile.

Women's higher emotional intelligence requires more subtle and sophisticated marketing strategies with them than with men.

Women and men tend to be better at regulating different emotions. Women tend to inhibit anger emotions, which is not surprising given their motivation for building relationships (Brody and Hall 2000; Fischer and Dube 2005). Their motivation to build relationships overrides their anger while the object of their anger is present, but they may vent this when alone or when the other person is no longer present. Men tend to inhibit expressions of fear and sadness as they deem these to be a sign of weakness based on their motivation for power. Women generally score higher on the Myers–Briggs personality test on the trait of neuroticism. This means that their emotional traits are higher, and that manifests itself as greater swings in emotions than men are subject to. This is a familiar stereotype, with women described as highly emotional, but this is only viewed as negative in a masculine, rational world. In a world that prioritizes emotional intelligence this should be an advantage.

HORMONAL DIFFERENCES

Estrogen and testosterone

The primary sexual hormones of estrogen and testosterone play a huge part in defining the feminine and masculine behavior of everyone. During the Seventies it was thought that the balance between estrogen and testosterone was a zero-sum game, that is an increase in one meant a corresponding decrease in the other. However, scientists now believe these are independent. There is a level of the estrogen hormone and one of testosterone that can be increased or reduced. As these changes occur, the high hormone level of one masks some of the activity of the other rather than negating it (see Figure 1.7).

A woman's cognitive abilities change over the time of her menstrual cycle based on the level of testosterone in her body, while a man's level of aggression changes through the circadian cycle, over the time of day and the seasons of the year, based on his level of testosterone. For women, a lower estrogen level is reflected in better spatial abilities, while higher levels (just before menstruation) are reflected in finer manual and verbal skills. A recent research study in Germany (Stallwood 2005) analyzed prenatal exposure to testosterone and its link to spatial abilities. It has long been known that absorbing high levels of testosterone (for men or women) in the womb leads to a ring finger that is as long as the index finger. This has often been used as a visual indicator of femininity or masculinity. The German

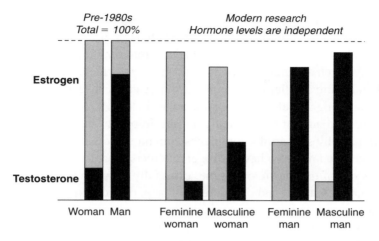

Figure 1.7 Sexual hormone composition that influences levels of femininity.

study tested people's ability to rotate complex shapes in their head and manipulate complex shapes. The differences in hormones between women and men have some interesting effects on their lives:

- After consuming the same amount of alcohol, women have a higher blood-alcohol count than men, even when you allow for size differences.
- Women who smoke are more likely to develop lung cancer than men who smoke the same number of cigarettes.
- Women have stronger immune systems than men, but are more likely to get autoimmune diseases.
- Depression is two or three times more common in women than in men, in part because women's brains make less of the hormone serotonin.
- Women's brains are less vulnerable than men's brains to aging effects, including less decrease in brain tissue mass.

The feminization of society is not just a brain or cultural phenomenon; it is a powerful physiological force that is affecting men. This change is driven by increased female hormone levels in the drinking water of many Western countries. There are several reasons for this as estrogen hormone traces from the contraceptive pill that are water-bound are not being removed during standard water cleansing and recycling processes. This means that as the water in a city gets cleaned and reused several times, the female hormone levels remain and are concentrated each time the water is used. Many studies have found river fish that are physiologically female but when their DNA is tested are found to be male. Similarly, other waterborne animals have been researched and there is a growing evidence of male animals and fish having smaller male genitalia. These cases have all been linked to abnormally high levels of female hormones in their bodies. Modern chemicals are also adding to the changing physiology of humans. The plastic bottles used by soft drinks companies contain a chemical called phthalate which is now known to have estrogenic effects on animals. The phthalate leeches out of the plastic over time, especially in warm weather, and into the liquid it contains, and therefore into the person drinking the soft drink, and thence into the water table. Several studies have noted that there is a remarkable congruity to the continuing drop in male sperm count and industrialized cities and countries. Since 1960, the sperm count of the average male has dropped by over 50 percent and infertility rates have quadrupled. Environmental estrogenic consequences are also affecting women. In the US, girls are reaching puberty earlier and earlier. Once considered abnormal, girls reaching premature puberty at the age of eight are now a regular phenomenon. The issue was considered so significant that *Time* magazine placed it on its front cover in 2000.

> Higher levels of estrogen produced just before menstruation increase women's verbal capabilities while lower levels reduce their spatial abilities throughout the rest of the month. Lower levels of serotonin in women lead to more cautious behavior.

Serotonin

Serotonin is often known as "the happy hormone" because high levels create a feeling of happiness and low levels increase mood depression in women. Its production is increased with exposure to sunlight and lack of it contributes to seasonal affective disorder (SAD). It also plays a key role in helping the brain make neural connections and therefore has the ability to change how your brand is perceived by women. Women and men's behavior changes differently when serotonin levels are reduced. A reduction in serotonin in women makes them more subdued and cautious, while in men a reduction makes them more impulsive (Leibenluft 1998). This has a direct impact on how we create brand experiences and adapt our service behaviors during the time of day and time of year. Increased caution in women means they will require greater levels of detail, more product research and a slower pace of service. This may have a higher impact in countries that suffer from lower sunlight levels for many months of the year. The reality is that women tend to do far more product research prior to purchase than men do. They will have compared many variables of two different brands before making a final decision. Men are more likely to be quickly swayed by an attractive headline and then buy the product without further research. While this is not conclusive, it shows that a more considered approach to selling to women will be more successful. It also indicates that rushing women to purchase is the worst thing that a salesperson can do.

Long-term brand relationships not one-night stands

Women are more interested in building long-lasting relationships, while men are more associated with satiation, consuming something and then moving on afterwards. They are less concerned that they may have many short relationships, rather than long enduring ones. This reflects one of the essential differences between men and women. A woman needs to find a steady mate to procreate with, one she believes will stick around for the long term and protect them and their offspring as a family unit. They also invest far more time, effort and emotion in pregnancy and rearing their child, while men invest relatively little in the pregnancy and, in primeval terms, need to mate with as many different women to ensure the success of their DNA. While this insight is anthropological it's also a great metaphor for women's and men's branded relationships as well.

CREATING POWERFUL BRANDED MEMORIES

Creating branded memories is the foremost means marketers use to change a one-off transaction into a long and loyal customer relationship. It is also pivotal in helping women to assess brand trial opportunities of new brands or brands new to them as a customer. Branded memories are also used to manage the level of emotional "tagging" that we attach to the brands we consume. Clearly, the memory is the key repository for creating and managing brand value in the mind of the customer. When marketers talk about repositioning their brand in the mind of the customer they mean in the memory of the customer. Without strongly branded memories, customers would act like goldfish, forgetting all they ever knew about your brand every three seconds. Accessing and helping to program the customer's memory are the key activities marketers should focus on in developing relationships with their customers.

The brain makes sense of all the sensory inputs we continually receive to create memories. It files them in a kind of "card index" so that if they are powerfully positive they will be remembered with strong emotional recall. We can test this by recalling a positive memory – a first date, a delicious meal – and notice that even the memory can trigger an emotional response in us many years after the original event occurred. These memories are categorized and subcategorized, initially into positive and negative; frightening memories can trigger just as powerful an emotional response as positive ones. The limbic system operates as the navigator to that index, deciding which memories go where and whether memories are superseded or replaced. Brands need to use emotional connectivity to increase the likelihood of strongly indexed memories.

This memory index system is based on a huge number of mental representations. These may be verbal, visual, experiential and sensorial. Neuropsychologically speaking, memories are created through a series of patterns of neurons firing simultaneously. They form a "fingerprint" that signifies the mental representation of that specific memory. Obviously a grown adult has many millions of these and each one is unique, allowing us to remember events or feelings from even decades ago. MRI scans of the brain can illustrate these patterns, which are extraordinarily beautiful. As Dr Harry Alder (1994) describes it, "The number of potential neural connections in the brain is greater than the estimated total number of atoms in the universe" – there is a staggering potential for rich memories.

> To create powerful brand perceptions brands need to create deeper brain engagement by linking their brand message with the woman's gender-based life goals.

Brands are a powerful shorthand to help navigate and reinforce memories within the index system described above. Our previous experience of an airline like Virgin or United Airlines gives us a memory bank of knowledge to judge new brand experiences against. When trying out a new service like Southwest or easyJet, we are able to benchmark and therefore make

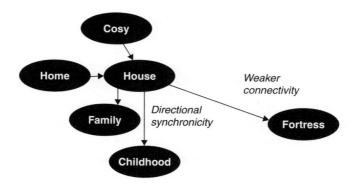

Figure 1.8 Simple mind map.

sophisticated brand choices each time we add another experience to our memory. Conversely, when we are faced with new categories, we rely on more famous brands to help us. So if we haven't flown before then we may choose a bigger, better-known airline like British Airways rather than a smaller start-up airline. This is precisely why brand extensions are a low-risk proposition for businesses. When we see that Nivea has created a range of skin products for children, we link that to the already known and respected women's range and invest that assurance into the new product. Without that prior brand memory we would have a much lower level of trust in the new brand extension.

Mind maps

The enormous numbers of memories we have are connected through mind maps or shared ideas and thoughts (see Figure 1.8). This allows us to quickly navigate what would otherwise be just memory "soup." Rational thoughts are more easily designated to different parts of any mind map. Large objects can be placed in order of size; numerical information can be placed in order of size of number. Dates can be "filed" chronologically; animals and plants can be organized by genus and type. The use of bucketing helps to keep the total number of discrete memories down and makes them all easier to access. Emotional or complex memories are equally stored within the mind maps but their location may be more quizzical and less obvious, though no less vivid or retrievable. In fact, by definition, the more vivid and emotional the memory, the easier it is to retrieve it. If we really want to make our brand more memorable, we need to use more emotionally

charged marketing content. This clearly favors women who are better able
to receive and store emotional messages and ideas.

Building emotionally charged memories

Our emotional state plays a key part in how our memories are assimilated
by our brains. If we are in a positive mood we are more likely to encode
memories that are positive or encode memories in a more positive light.
Likewise we are more likely to retrieve positive memories or retrieve them
under a positive "spin" when we are in a positive mood or experiencing positive emotions (Schacter and Scarry 2000). The same is conversely true for
negative emotional states and the encoding and retrieval of negative memories. Making women feel good during their brand experience is one of the
most effective ways to increase these branded memories; this is described
in more detail later in this book with the feelgood framework. We are able
to increase the power of our memories by deeper engagement with the
subject-matter. So for example if we superficially read an advert about a
car that has three features, we may remember the car brand and one or
two of the features. However, if we read and analyze the features engaging
our brains we are able to create deeper memories because there are more
neural connection patterns about the brand. This might be through understanding the impact of the features on our driving ability, characterizing
what might be different about the experience from our own car, mentally
placing ourselves in the driving experience. All these additional brain iter-
ations ensure that we are more likely to create stronger memories for that
particular brand. It may sound simple, but creating emotional engagement
by making women feel good about themselves is the best way to creating
a stronger brand impression.

In order to keep a brand at top of our mind, we must place the branded
emotion into our long-term memory. We can often recall exquisitely
detailed memories from our childhood, even in old age. Long-term memory
is virtually limitless and lasts a lifetime. The process of recalling long-term
memories involves transferring them into the short-term memory in order
that they can be brought into "consciousnesses." People do not remember
a list of things equally even in long-term memory. There are two psychological effects that take place. The first is the primacy effect, with the first
things on the list creating a strong impression. The second is the recency
effect, where the last things on the list are remembered the most strongly
(see Figure 1.9).

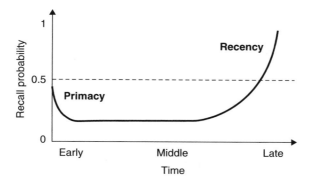

Figure 1.9 Primacy and recency effects in memory recall.

Women have better brand memories than men

Women have better brand memories than men because they are better at emotional interpretation and recall than men. As we all know, women can often remember the tiniest of details about what they were wearing on their first date, while men struggle to remember even the time of year of that crucial event (Brizendine 2007). In real life, our memory is managed not in isolation but is influenced by our own outlook and motivations in life. There are also implicit gender predispositions in the way we remember brands. Before customers will remember a brand, we must first ensure it has high perceived relevance to their life goals. Otherwise they will view the brand as insignificant and apply less memory power to our brand. This means communicating the relevant big-picture benefit to women before they are willing to engage on the specific relevance of a particular brand. The supermarket Tesco illustrates the overall benefit of thriftiness and service (key grocery shopping needs) with its "Every little helps" slogan so that women can begin to connect with it as a brand. Herrmann *et al.* (1992) conducted an experiment to demonstrate the effect that different gender-based life goals had on memory performance. The results showed that men demonstrated a dramatically different performance between those memory tasks that were clearly linked to their life goals and those that were not. Women also showed a difference but a much less significant one than the men. The men remembered almost a third less items that were not related to their masculine life goals, while women remembered only slightly fewer items that were not related to their feminine life goals than were related

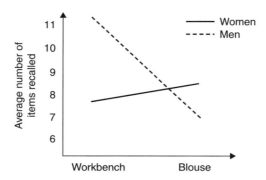

Figure 1.10 Women's and men's performances recalling items not related to life goals (Source: Westen 2002).

(see Figure 1.10). This means that it is easier for marketers to get women to remember their brand image than men.

Strengthening brand perceptions and memories requires you to connect with women. The more they engage and have to "process" the brand experience, the stronger the emotional memory of your brand. You need to avoid creating light brand memories where a woman only remembers the color of a new bread brand package rather than the emotional benefit of a specific brand like Hovis. You can increase the strength of the brand memory by emphasizing specific and distinctive details. For example, the supermarket shopper may analyze the package and see that the brand contains only organic ingredients. This encourages the woman to replay the slogan and brand image in her head and link that to a positive brand perception. Finally, you can ensure deeper brand memories by emphasizing the emotional benefit of a feature. Continuing the example, the supermarket shopper interprets the brand and its organic ingredients to mean healthier food for their children. This richer emotional territory links powerfully to a core emotional value of the woman and is more likely to be positively perceived and therefore purchased. As we can see, the more we can link to the emotional meaning for each individual customer, the more likely they are to have a more powerful memory of our brand.

Brands that trigger active remembering through increased brain engagement (where the brain replays imagery and feelings) result in stronger brand memories that keep your brand top of mind.

Media planning frequency to help build memories

The length of the cycle of repeated stimuli to create memories also has an influence on the strength of memories and the memory degradation curve (which indicates amount of memory lost over time). This is crucial for advertisers, who try to build up a long-term brand impression through repetitive advertising spots. For an initial brand launch, a flood of integrated print, radio and television adverts can create an initial bow-wave of awareness and interest. However, rather counterintuitively, Bruce and Bahrick (1992) found that the spacing of stimuli sessions at longer intervals created deeper and stronger memories. In fact, they concluded that lengthening the interval by two or three times tends to double the retention of the information. It is therefore unnecessary for advertisers to constantly bombard women with adverts, but they should define the appropriate input cycle (making it longer rather than shorter) in order to build powerful memories about their brand. The same effect is true for internal communications campaigns or brand launches. Again, an initial launch to generate high emotion and awareness can be very effective. But in the long run a slower drip feed with longer input cycles will help to increase an employee's retention of internal brand messages.

INCREASING BRAND LOYALTY WITH MEMORIES

As we have seen earlier in this chapter, memories are defined through a series of neuron patterns in our brain. In order to make these patterns connect more strongly with your brand they need to be active or "sticky." Rather than just ensuring that women read an advertisement, we need to engage them to "play" with the information in their mind. Women, with their higher verbal and comprehension abilities, are better able to "play" with communications and therefore should find it easier to memorize advertising messages and communications. These actions are called maintenance rehearsal (passive remembering) and elaborate rehearsal (active remembering). There are a variety of marketing techniques for achieving this. One is to create a psychological "completion" exercise by leaving a message or image slightly "unfinished" so that the reader will mentally complete the image in her mind. This extra level of interaction both personalizes the message (each woman may complete the message slightly differently) and ensures that they actively engage with the message topic. The higher the level of engagement the stronger the memory will be (both positive and negative memories). This is one of the reasons why brands try to develop brand experiences rather than just messages. The message will be received, but may fail to create a powerful memory because it has not engaged the woman on any deep level, but the brand experience engages women through a variety of sensorial stimuli, ensuring that deeper level of engagement is created.

SUMMARY

As we have seen, there are fundamental biological and hardwired brain reasons why women and men have different brand perceptions and purchasing behaviors. The brain and central nervous systems of women and men are different in specific ways that provide women with greater abilities in language, communication, emotional intelligence and empathy and relationship-building.

Hormonal differences between women and men mean that women are more group-oriented and less egocentric and achievement-driven. Marketing strategies need to align with these differences to be successful.

Women are better at articulating their emotions and the emotional relationships they have with others. They are more adept at subjugating their needs and show greater empathic capabilities than men.

They have greater language and listening skills which means they are better at creating and understanding marketing messages.

Women build stronger relationships with their favored brands that emphasize the same feminine brand traits that they prioritize themselves. Their brains also make them more considered in their judgements and responses to marketing messages. This means that they are more likely to cross-reference previous memories with new experiences and brands and not make rushed purchase decisions.

2
Women's primary characteristics

In this chapter we explore and explain the primary attitudes and characteristics that all women possess. These are the fundamentals that underpin women's key brand choice and purchasing behaviors. Women's six primary characteristics are described in detail with examples of how they affect their purchasing behavior. We then go on to identify the marketing consequences of these behaviors and illustrate ways to maximize marketing impact with women.

OBJECTIVES

- Define the gendered behavior spectrum
- Identify what drives women's behavioral differences
- Identify female need states and why they are dynamic
- Identify the six primary characteristics of all women

THE FEMININE SPECTRUM

For the purposes of clarity, the differences between women and men have been described in this book as a binary opposition. However, the reality is that there are some women who are more masculine just as there are men who are more feminine. It is an overlapping continuum or spectrum that applies as much to men as to women. It is often said that the "new man" is in touch with his feminine side and this is evidenced through different audience descriptions like "the metrosexual male." This reflects such people's gendered attitude and behavioral orientation rather than their sexual orientation. It is critical that you define the level of femininity of your target women rather than treating them simply as a single mindset. Without this, any gender-based marketing will be too superficial, crude and unsuccessful.

> When marketing to women it is crucial to avoid binary stereotypes of "women and men." Gender-based attitudes of femininity and masculinity are not polarized but spread across a spectrum of shades of difference. You need to be specific about the levels of femininity your target woman requires within your market sector.

Women are highly adaptive, and demonstrate this through their multitasking abilities. This means that the balance between the need to make use of feminine or masculine traits is likely to vary depending on the role or task they are completing. Each female segment has a "center of gravity" along the spectrum of femininity. The range of that variance is likely to be higher for women than for men because of their wider emotional range. Men typically have limited emotional and empathizing abilities, meaning they are less adaptive to others and external situations. Women use their greater perceptive and empathizing abilities to comprehend and modify their attitudes and behavior to individual situations. Their multitasking lifestyles also give them far greater experience of this type of continual adjustment and over time they build up a greater repertoire of different balances or submodes within their own feminine behavior (see Figure 2.1).

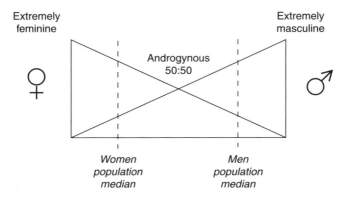

Figure 2.1 Women's and men's median values on the feminine–masculine spectrum.

This continuum will affect marketing on every level. The rest of this chapter describes only the direct opposites between women and men to highlight each point more clearly.

WHY WOMEN BEHAVE DIFFERENTLY

There are several theories as to what defines these gendered behavioral differences between women and men. Most scientists agree that there is a complex connection between biological and social factors, but here is a summary of the key influences on the gender debate. By understanding the main theories about how gendered behaviors are created, we can identify key marketing strategies that focus on psychological and personal or social and group-oriented messages.

Biological differences

Biological theorists argue that there are genetic differences between the sexes; women are weaker than men and produce reproductive eggs not sperm. These cause each sex to take up roles more suited to their biological makeup. The women reproduce, while the men hunt and dominate. These tasks require each to develop higher abilities for their task, so men are more aggressive and have higher spatial ability while women gain nurturing and social abilities. Biological theories rely on there being little impact by a changing society to influence the differences between women and men. Marketing strategies that emphasize this rely on time-honored imagery of the feminine wife, mother and home-maker. Brands like Fairy Liquid, Persil and Dixan use this basic semiotic imagery to target women, while men are depicted as powerful fighters conquering evil, with brands like Omo, Domestos and Cillit Bang.

Social learning differences

The personal development concepts of conditioning are the core tenets of social theory. Conditioning of children is achieved by rewards and punishments and modeling through observation and copying. Children learn and copy gender-specific behaviors from their parents, siblings and friends, plus television and other media. Parents therefore play a key role in this as they generally provide the reward and punishment for learned behavior.

Social learning theory recognizes that both women and men have the capacity and ability to learn either gender role equally, but this does not mean that they will behave correspondingly. Unlike other theorists, learning theorists also believe that learnt behaviors can be unlearnt through retraining. Brands that communicate rewards and punishment for on-gender behavior benefit from this type of behavioral learning. The L'Oréal shampoo campaign communicates the reward of being more feminine with shiny hair through its "Because You're Worth It" campaign, while the Marlboro cigarette brand challenges men to be a "Real Man" and smoke Marlboro, or accept that they may be too feminine for their cigarettes.

Psychological differences

Based on Freud's theories, psychoanalysts believe that gender differences emerge through a child's development during the ages of three to six years. A girl develops an inferiority complex around penis envy and therefore strives to gain favor with males throughout her life, while a boy rejects his mother because she appears to have been castrated and strives for acceptance by his father as he is obviously the dominant force. Clearly in the twenty-first century these ideas hold little credence. Crude or tongue-in-cheek advertising may still use clear visual symbols of masculine or feminine anatomy but by doing so it will turn women off completely.

Cognitive differences

This theory focuses on child development and the "stages" they grow through in terms of cognitive abilities. By the age of six or seven, children realize their gender is constant, as evidenced through self-categorization as "I am a boy" or "I'm a girl," and this leads them to develop stereotypical behaviors to reinforce this categorization. The section below on female life stages expands this in more detail. Brands like *Just Seventeen* magazine help young women to reinforce their life stage, while brands like Mothercare or Home Depot clearly identify a woman in a later life stage as a mother or home-maker.

Women's purchasing behaviors are based on a complex mixture of genetic and socially learned attitudes. Try and gain insights about the life stage and general upbringing of your target woman.

Some cognitive differences are whimsical, but collectively they contribute to the differences between women's and men's purchase behaviors. There are three issues that define our gender type and these can be categorized as follows:

- Genetic sex, the XX or XY chromosomes that make up our DNA, hormone balance and the type of sexual organs we possess. These are binary and birth-defined.
- Brain type, cognitive thinking, whether we are largely an empathizer or systemizer. These are influenced by social and cognitive influences.
- Typical gendered actions and behaviors, for example women chatting with each other about their feelings, while men play with gadgets and football. These are influenced mostly by social learning.

Some of these are fixed at birth while others are influenced by society and our upbringing. Table 2.1 presents an overview of the findings of extensive past research into feminine behavioral differences.

FEMININE NEED STATES

Need states are created when there is a schism or gap between our defined goals and the daily reality of our lives. These provide the driving force for our attitudes, behaviors and decision-making criteria during purchase moments. This schism or gap is often referred to as cognitive dissonance – the difference between a woman's desires and her reality. We examine the marketing opportunity that this creates later in the book. We each of us have a large, complex variety of needs that are constantly managed in terms of priority and action through our behavior.

In women's minds there is a general layering of need states that begins with their general value and belief systems. These can range from the needs for inclusion and acceptance through needs for healthiness and purity, peace of mind and stability, excitement and energy or status, glamour and beauty. It is important to recognize that these have a dynamic and context-bound character and rarely remain constant. This means that women have to constantly rebalance our priorities across the following dynamics:

- variety of needs throughout the day, week or year
- dynamic nature of needs

Table 2.1 Research findings into feminine behavioral differences (Source: Lippa 1994; reproduced by kind permission of the author).

Behavior (positive 'd' value denotes females higher)	Mean value of 'd'	Number of studies
Social behavior	0.43	64
Social smiling	0.63	15
Personal space	−0.56	17
Expansiveness of movements	−1.04	6
Filled pauses (ah's and um's) in speech	−1.19	6
Aggression	−0.50	69
Group conformity	0.28	35
Helping – overall	−0.34	99
Helping – while being watched	−0.74	16
Helping – while not being watched	−0.02	41
Behavior in small groups – Positive socio-emotional behaviors	0.59	17
Behavior in small groups – Task-oriented behaviors	0.59	10
Leadership – overall in lab groups	−0.32	74
Task leadership	−0.41	61
Social leadership	0.18	15
Democratic vs. autocratic style	0.22	23
Cognitive verbal ability	0.11	165
Cognitive mathematics ability	−0.43	16
Cognitive visual and spatial ability	−0.45	10

0.2 – small difference; 0.5 – medium difference; 0.8 – large difference.

- source of needs, internal and external
- prioritization of those needs

Clif Bar, the US candy firm, identified a new need state and connected with its customers around an emotionally and physically healthy lifestyle. Clif Bar recognized an opportunity to attract women to its products, launching the first energy bar created specifically for women, Luna. Clif Car developed the product according to women's unique nutritional needs, enlisting help from the company's female employees during flavor conception, tasting and naming. Leveraging a multimillion dollar budget, Luna bar's marketing involved concert tie-ins, book club events and an annual movie festival for women – LunaFest – with proceeds earmarked for the Breast Cancer Awareness Fund. The company promoted Luna bar using a spiritual message to resonate with its target audience, the tagline of which reads in part: "You, well fortified. The blissfully good, whole nutrition bar . . . to promote wellness." Luna Bar further connects with women via its website, which features community-based clubs, such as Luna Moms Clubs and Luna Teen Chix. It also encourages women to create a "Luna Life-ism collage" to capture what inspires them personally. The success of Luna Bar created a new female-targeted niche within the food and energy bar category. Luna enjoys a loyal customer base today.

The range of different needs can be quite enormous and our minds are constantly reviewing, ordering and prioritizing these at any one time. They are often very different in nature, scale and frequency. Research has shown that women's needs are more strongly focused on relationships with others. This means that brands that can enable and reinforce social networking and friendships like Facebook or Starbucks will always be favorable to women, while men prioritize task-oriented needs that have a clear completion. Marketers should develop a woman's "world view of needs" in order to identify new product service opportunities.

Primary and secondary characteristics

A woman's behavior is constantly adapting, whether at the secondary level of such matters as for example whether to have a cup of coffee or of tea, or at a more primary level over decisions regarding such things as starting a family or buying a house. The cause of these changes can be driven by either internal or external stimuli. The ability to flex and reprioritize between primary characteristics and secondary characteristics is a universal skill that all people possess. It is important for marketers to gain deep insights into the drivers of those behaviors in order to create effective marketing campaigns. This means going beyond the simplistic descriptions of customers as "middle class" to a more accurate understanding of their daily flows of needs and attitudes. There is a huge opportunity for marketers as they can disrupt this sequence and suggest or influence women's needs with suitable products. There is a defined primary set of characteristics that all women display. These are described below and underpin the core behaviors of women. In addition to these constant characteristics there are secondary characteristics that are more present in different types

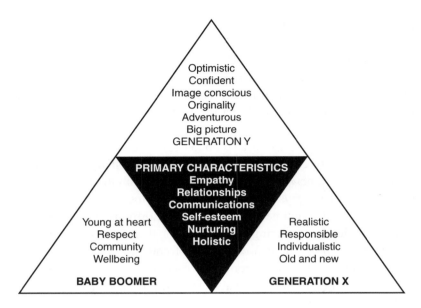

Figure 2.2 Primary and secondary characteristics in Generation X, Generation Y and Baby Boomer women.

of women. Chapters 3, 4 and 5 provide a detailed analysis of these differing secondary characteristics in Generation X, Generation Y and Baby Boomer women. The challenge for marketers is to understand the rules by which people prioritize their needs. They need to identify when and by how much their purchase choices are driven by the primary and the secondary characteristics of all women (see Figure 2.2).

EMOTIONAL EMPATHY

Higher emotional intelligence

Empathy quotient is an index of how well we are able to perceive others' feelings (especially beyond what they are telling us). Body language is one of the key ways women pick up on other people's feelings. Women are much better at picking up on non-verbal cues and understanding them and therefore can have greater empathy with other people. This is often described as women's intuition and can be derided as a myth. But the reality is that women's brains function with a less egocentric attitude, allowing them to perceive subtle differences in people's behavior and respond more appropriately to that signal. This is why women are inherently better at helping people than men. This may seem prejudiced, but it is also why they are better at dealing with others than men. That's not to say that all men are no good at this or that they can't learn – but they are starting from a much lower base than women. Empathy quotient also means the level to which we are affected by other people's feelings.

Women are both better at perceiving others' feelings and adapting themselves to those feelings. You need to demonstrate that you recognize and appreciate what women are feeling. Marks & Spencer the UK retailer has demonstrated a high level of empathy quotient with the implementation of "Plan A" because there is no "Plan B" set of initiatives. These cover a range of empathy exhibiting commitments to environmental standards, workers' rights and partnerships with the charity Breast Care Awareness. The combination illustrates how a large corporation can demonstrate that it cares about others and this provides a strong message to female shoppers that this is the ideal place for them to shop.

Women have a much higher emotional intelligence than men. This means that they are more perceptive, especially during conversation. This enables them to build more powerful and longer-lasting relationships. Women also have a lower threshold of risk than men. This can easily be evidenced by the far lower incidence of driving accidents in young women drivers. Insurance companies routinely give women drivers lower insurance premiums as a result of this behavior, and women-only brands like Sheila's Wheels focus entirely on this benefit. Recent decades have seen a significant rise in both women and men externalizing their emotions. Language is the channel by which many of our emotions are communicated, along with nonverbal forms of communications. Seemingly superficial changes in language usage are a concrete sign that cultures have embraced deeper philosophical changes. Self-help books like *Computers for Dummies* or *Who Moved My Cheese?* have moved from the niche novelties to the mainstream brands that cover a wide range of products and services. They are effective because they demonstrate how to use simplicity, storytelling and self-deprecation to gain acceptance with women audiences (see Figure 2.3).

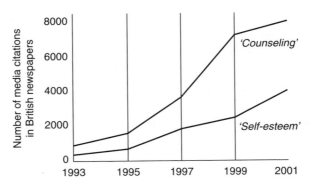

Figure 2.3 Recent rise in emotional language in everyday usage (Source: Furedi, Frank, in the *Metro* newspaper, London, 15 March 2001).

High intimacy counts

Intimacy and personal details are the topic of women's conversations more than men's. For example, a female friend of one of the authors described how she had seen a group of women in the office discussing the latest underwear they had bought, each contributing their point of view on the product and explaining why it was more comfortable. She pointed out that men rarely have this kind of conversation; it is just too personal and intimate. Women like to know all the details, even if this requires divulging highly intimate knowledge, while men tend to avoid all unnecessary intimacy. Women see intimacy as a way of building a higher degree of emotional support than men within their relationships. It becomes a journey and a metaphor for the state of their relationship with another woman. Brands that can tap into higher levels of intimacy will gain strong loyalty with women. Agent Provocateur, the lingerie brand, has achieved higher levels of overt intimacy by articulating and legitimizing women's desire to look and feel sexy for themselves, as opposed to just for their man. Their shops, their website and their "Adventures in Lingerie" series of stories starring actress Maggie Gyllenhaal all encourage sharing of intimate details as a modern and cool brand.

Women have higher emotional empathy than men. They seek and appreciate higher levels of dialogue, social benefit and intimacy from their brands.

Women have a much stronger sense of romantic imagination that is accessed through escapism and fantasy. Their mind is a powerful image-maker in its own right. The extremely vivid nature of dreams, daydreams and inner thoughts can often trigger physical responses. The imagination creates a world so vivid that the body responds to it with a chemical reaction just as it would to external triggers. TV brands like *Sex in the City* provide women with a version of lifestyle of play with their daily existence, romance, relationships, entertainment and dining. Tiffany's the luxury jewelers reinforces the romance that women should expect and deserve with seductive advertising for everything from engagement to eternity rings. Like a dream, they often combine disparate areas of life in ways that are imaginative and manage to create a coherent narrative. There are several recognizable genres for the imagination, and identifying the most suitable is important in developing a clear marketing approach to women. The most effective are those that align more strongly with the typical female traits and themes:

- happiness
- romance
- emotional situations
- self-improvement
- self-actualization

(These are more deeply defined in the "feelgood framework" later in the book.) Fantasy and imagination can play out into reality. This may be through make-believe games in childhood or acting on stage as an adult. The act of play is something that starts as the chief component of daily life as a child. While the time spent on play is greatly reduced as adults, the urge and need to play is not. Women may use more sophisticated forms of play but underneath the key is to unlock the child inside through external and physical activities. Theme parks were the first to recognize this and cater as much for adults as children. Successful shopping malls and stores have incorporated elements of play into their shopping experience. This makes them more enjoyable and memorable, encouraging the shopper to return. Virgin Atlantic's social "Party In The Sky" brand attitude encourages more women to fly with them than the more traditional British Airways. It offers an adult version of play where life is still fun and the romance of travel has not been lost.

VALUABLE RELATIONSHIPS

Women and men have distinctively different attitudes towards relationships, a characteristic that is one of the most significant differences between the sexes. The differences are also the most important for marketers to understand if they are to market successfully to women. As Emma Laney (2005) puts it, great brands create ties and bonds of an intimacy similar to that found in female relationships.

Women show more positive social–emotional behaviors than men by:

- engaging with friends
- agreeing with others
- offering emotional support

Men show more task-oriented behaviors by:

- giving and asking opinions
- solving problems
- searching for resolution

This means that women are more effective in group tasks requiring discussion and negotiation. They prioritize and value relationships far more highly than men do. These are characterized as benevolent and reciprocal, where each person stands to gain equally: "Women have the emotional capital to develop and keep friendships and support networks, whereas men tend to become more isolated" (Scase 2000). Men on the whole prefer work relationships that have a clear power structure and are based more on competition. To characterize the difference, women prefer cooperation while men prefer competition. Women also show stronger concern about maintaining the status of a friendship and worry more about losing a friendship. Their focus of effort is therefore on communicating and nurturing close relationships. Even if there is not a specific task or activity in hand, it is important to invest in the maintenance and development of the relationship itself. Women especially prioritize intimacy, closeness, shared secrets and eye contact. Starbucks is a very female-friendly brand, with its ethos of socializing and inclusivity that encourages women to bond with their brand. This is in stark contrast to the more machismo-driven, male-oriented Italian coffee houses like Café Nero and Costa Coffee that focus on the semiotics of power and technical mastery of machinery to attract male coffee-drinkers.

Joining a club

Women have a strong need for group membership with others. The desire to be included is driven by their fear of being excluded and therefore solitary and lonely. Women see the action of exclusion as the demonstration of nonacceptance by others. To be included is equally a demonstration of acceptance. This is not a fixed and binary relationship but one that is dynamic and changeable. A large part of childhood upbringing is used to understand and develop the rules and mechanisms to manage the process of group membership. When a 6-year-old girl says she has a best friend, this may be one of many "best friends" or the friend in mind may change regularly as patronage is shifted. This simple anecdote describes the basis for an entire adult social network. The driving need for membership is part of modern culture. Beyond the functional need for survival through group safety, one of the main emotional needs is to associate and be associated with others. Gwynn Burr, Director of Customer Service at Sainsbury's, suggests that "Women prefer to be a member of the winning team, rather than leader of the losing team." So if the choice is between being a poor leader or a supporting team-member, they make a more effective tradeoff than men. Brands play a crucial part in delineating social groups, including those that are leading or cool. They can help them both to choose and to signify their inclusion or membership within these subgroups. Brands like Chanel and Prada are perennial female cool brands; being a member marks a woman out as one that is in the know.

Women will be more attracted to brands that enable and promote inclusivity, relationships and social esteem than to those based on achievement alone.

Independent Asians

Women's emancipation in the West has made strong progress over the past century and is also beginning to be witnessed in Asian countries as well. Central to this equality is the level of independence that Asian women desire and attain in modern society. Social freedom is born from an increased economic autonomy that Asian women are now achieving. Countries like Japan, China and Korea have all advanced beyond pure manufacturing and are now well-developed service economies. This has increased opportunities for Asian women who are adept at these tasks and can often also work from home. Being able to earn their own living without relying on their husbands is a crucial milestone in Asian women's sense of independence. This means that they now have disposable incomes to spend freely on personal and fashion items. This financial stability has led many Asian women to get married later in life. Singaporean women are now getting married on average one year later than they were ten years ago (Wee Guan and Chew Su-Fern 2005). Big brands are waking up to this nascent market opportunity and the life assurance brand Prudential has recently launched "PruSmart Lady," a series of life assurance products directed specifically at women (prudential.com.sg 2007). Similarly, the Hong Kong financial services giant Fubon Bank has launched a series of credit cards targeting women by partnering with Bonluxe, the exclusive Asian lingerie brand. The card not only provides a competitive rate but gives these women additional shopping discounts and privileges at Bonluxe stores as well.

NURTURING

Women have a strong need to establish and nurture relationships with others. This satisfies their need for both giving and receiving care, affection and emotional support. It also provides reassuring feedback from others that reinforces their confidence and sense of self-worth. Nurturing and affection demonstrates goodwill to others, something women find much easier than men. The amount of affection shown is often used by others to gauge and manage the closeness of their relationships. There are many ways women can show affection, whether this is being considerate to other train-users by giving up their seat or giving presents or helping friends. Women have a strong social and psychological need to develop friendships, to be with those they like and enjoy doing things with, either passively like going to a movie or actively like playing sport. Women tend to be more driven to achieve stronger and longer-lasting friendships than men. There are several rules for those involved in female friendship:

- they help freely in times of need
- they trust each other
- they respect each other
- they share confidences while respecting each other's privacy
- they expect regular friendship maintenance
- they are interested in stories rather than facts

Women are more willing to invest in their brand relationships and those brands that enable socially supportive activities.

Ford encountered difficulty targeting women owing to the increasing fragmentation of its traditional marketing channels. Seeking a new method to attract this segment, Ford turned to cause-related marketing, committing $60 million to the Susan G. Komen Breast Cancer Foundation Race For The Cure. Ford's philanthropic efforts included the creation of a unique Kate Spade silk scarf auctioned at Bloomingdale's, involvement of more than 4000 Ford dealers as local sponsors at 112 Race For The Cure events, and distribution of more than a million Breast Cancer Awareness bandanas to race participants. By highlighting its nurturing nature, the Ford brand enjoyed a huge boost in popularity among women and a significant increase in sales.

Friendship circles

The need for emotional support satisfies women's need to be social rather than alone. Certainly in modern culture there is a view that prioritizes being part of a social group, especially the family unit over being alone. Being alone is often characterized as strange, antisocial and mildly unacceptable. Women are much better at developing and maintaining social groups precisely because they are much better at nurturing and socially supporting each other. MSN recently reported that in April 2007 after eBay the most popular website was the social networking site Bebo. Similar sites such as Facebook and MySpace were also in the top ten. Overall, the internet in the UK was used by more women than men in the age group 25 to 49. Of total internet use in the 25 to 34 age range 55 percent is female. The rise of such social networking sites has also encouraged women internet users because it plays to their strengths of social nurturing and bonding far more than it appeals to men. In modern society one of the most widely recognized benefits from satisfying emotional support is the reduction in the stress which now seems omnipresent. Stress is the curse of modern society and affects an extraordinary number of people through work and family life. People's need for nurturing increases dramatically during stressful situations. Emotional support acts as a buffering tool and helps to shore up people's behavior and demeanor in the same way a massage often releases tension in one's neck or back. Unfortunately this simply alleviates the symptoms rather than solving the problem; removing the cause or causes of the underlying stress would be far more effective.

Loews Cineplex provides learning, socializing and enjoyment opportunities aimed specifically at young mothers. In fall 2002, Loews Cineplex began hosting Tuesday morning movie screenings designed for stay-at-home parents and their babies. The promotion, called Reel Moms, proves an attractive option for young mothers who crave more social interaction with their peers, an improved awareness of popular entertainment, and an opportunity to watch a movie in a theatre where noise is permitted (and common). Loews expanded the promotion across major US cities based on its success.

SELF-ESTEEM

Tribal branding

Women prioritize group-belonging more than do men, who tend to prioritize individualism. The need for belonging is a key opportunity for branding and brands. Visibly joining a tribe through the active representation of a company's logo or identity is an important goal for many different groups in society. While this may seem strange it links to people's increasing need for identification and navigation. As society becomes more fragmented into smaller quasi-tribal groups, people need greater help and easier signposts to help them navigate the choices they must make. Clearer tribal identification can often come in the form of explicitly branded goods – a Burberry check scarf or Ralph Lauren embroidered T-shirt. Women also have brand needs because brands have become part of their goal systems. For example, a woman may have identified her goal of becoming successful and come to view certain brands as symbols of that success. But these act in a dialectic as the symbols themselves can often reinforce the perception that she is successful and thus make her more likely to succeed. Another form of brand goal is to become part of a social group, tribe or club. These are often demarcated through the use of brand symbols; therefore the need to own these branded goods and be seen at branded venues helps satisfy these needs. Apple and its range of iPod players have persuaded millions of women to become more engaged with consumer electronics, once the bastion of male exclusivity. The iPod's androgynous aesthetic, intuitive ease of use and helpful retail staff have all changed women's perceptions of consumer electronics and encouraged them to proudly join the iPod club.

Using brands as "clubs" is a highly effective and specific form of consumer demand generation. The art of brand-building has become so successful that the substrate is often less valuable or desirable than the applied branding. This is true for Coca-Cola, which often fails in blind taste tests, but is still preferred by consumers over Pepsi-Cola. Again, a Calvin Klein T-shirt is only fractionally different from supermarket generics but commands and receives a huge price differential based on the branding alone. The demand is for the brand, its semiotic meaning and a statement of conspicuous consumption rather than simply the product or service. Women's brand needs can be divided into two major categories, functional ones and emotional ones. Each of these can be exhibited as social or psychological attitudes or behaviors. The socialized needs relate to their outer-directed, external world, through the interaction with others, while the psychological needs relate to the inner-directed, personal and mental world.

Increased self-perception

Women have a strong psychological need to fulfil their sense of confidence and self-esteem. Brands can appeal to these needs by demonstrating their role as an enabler of that confidence. Women are very good at hiding their lack of confidence, so marketers need to find subtle and alternate ways to communicate with women that does not speak directly to this need. Understanding the root cause of a lack of confidence will help define the most effective approach. It may be that a woman feels overweight, or unattractive. The recent Dove soap campaign has used a range of normal-sized, normal-shaped real women rather than the supermodel waifs who often appear in cosmetics promotions. By doing this, Dove helps to send the message that women should be confident about themselves and comfortable with their shape and size. The Dove brand therefore became associated with something that helps to build a woman's confidence as much as with cleansing the skin.

> Women greatly appreciate brands that help strengthen and maintain self-esteem through confidence and strong brand status.

Women's symbolic needs relate to their perceptions of themselves and how others perceive them. Many fashion, cosmetics and car brands focus on the symbolic needs of the purchaser. For example research has shown that Saab drivers tend to be inner-directed and focus on the symbolic non-social need; they want to perceive themselves as successful. By contrast, BMW drivers tend to focus on the symbolic social need; they want others to perceive that they are successful. Women's emotional and symbolic needs have expanded in response to the reduction in the difficulty of satisfying our functional needs. As modern society has become more mediated with a proliferation of ideas and images, people have been able to articulate and define their symbolic needs in ever greater richness and texture. In some cases these have on the face of it become absurd, with adverts attempting to solve a symbolic need for softer toilet tissue to demonstrate that the owner is a better host because they have flower-embossed toilet rolls. While this may have some vague functional difference, it is certainly a symbolic social one. Many brands focus on the symbolic need for status, belonging and achievement. This is particularly prevalent in the fashion industry, which can deliver against all three of these needs. Wearing an Emporio Armani skirt provides easy evidence of status and the ability to pay for expensive skirts. Through doing so it establishes that the wearer is successful and has achieved a wealthy state. Finally, it is one of the most obvious badges of a select club. Other wearers assume a level of connectedness and a similarity of attitudes and behaviors, simply from the badge they wear on their skirt. While women are less competitive than men they are not immune to status symbols. Brands are exceptionally suited to helping women define and negotiate the highly complex landscape of personal differences. It may not be too presumptuous to say that in modern society women are now what they buy. The need for self-identification and belonging is huge and easily satisfied by branded goods and clothing. Increasingly, the granular language of brands and sub-brands helps to differentiate group members from the wider population through extreme symbols that evoke tribal connotations – for example, the wearing of highly exclusive Alexander McQueen designer dresses, of which only two or three are produced for the entire world. Limited production is the best way to ensure that women can be different and stand out from the crowd. The need for such a coordinated self-image reflects a greater level of need or obsession for a particular status and requires a commensurate level of commitment. For example many customers may buy their whole lifestyle or self-image from a single source. Fans of the Ralph Lauren brand often buy everything from his branded clothes, through bed linen and house paint, to a pair of sunglasses. This suggests considerable insecurity and a strong need for status even in the choice of paint brand.

Stimulate my mind

By the time women become adults they have become extremely sophisticated communicators with an appreciation for art and visual images or taste of food and wine. This mental activity behaves like a muscle and needs to be exercised and stimulated to satisfy their needs. Marketing plays a strong part in delivering this sensory stimulation whether it is through targeting products and services or providing witty advertisements on television for them to consume. All forms of media are particularly good at satisfying a woman's hedonic desires. But the need for cognitive stimulation can be something as simple as gaining mental resolution from completing a task well or it may come from learning a new language or a new way of working. Cognitive needs can be numeric, like enjoying using mental arithmetic in a store to cost items. They can also be linguistic needs, and conversation is the chief satisfier of those needs. Clearly women's greater linguistic ability means that they require not only greater linguistic stimulation than men but also a wider, deeper and more frequent set of experiences to provide it.

COMMUNICATION EXPERTS

Word-of-mouth specialists

Women rely on and use "word of mouth" knowledge much more than men. This may seem another way of saying that they like to gossip; but this is far from the case. Women are more likely to trust their friend's opinion on an issue or choice of brand than an advertisement or the advice of the salesperson. They each see the other as a trusted source and one that has their best interests at heart. It is an equal relationship rather than one where information is withheld to engineer power differentials, as typically happens between competitive males. Group membership provides women with a range of trusted sources on any issue. This may be as prosaic as helping decide which shampoo to use, or the benefits of a vacation destination. Women seek and are always open to additional advice or opinion. The stereotypical male, in contrast, is characterized as one who will not ask for directions, even when lost on a journey. Men find it very difficult to expose their lack of expert knowledge, as this is perceived as a sign of weakness or loss of competitive advantage. Women especially like to compare and buy brands that others use in a similar situation and are therefore "guaranteed" to fit it. For example, a woman who has been on vacation with a specific hotel brand is more likely to be approached by someone who also uses that hotel brand. The same is true for buying complex financial branded products like mortgages and pensions; women like to compare what their trusted friends have done in similar situations. This is particularly true when the purchase is new, complex and relatively infrequent (like buying a house). Women recognize that they will rarely build up sufficient experience with a specific realtor to make them experts, so they rely on external comparative support. The use of accreditation can make a huge difference in reassuring women about the validity of their choice both pre- and post-purchase. For example, a toothpaste brand accredited by a national dental association provides an objective social comparison for women.

Women like to affiliate with brands that have a greater knowledge of a situation than them and represent themselves as expert in the field. This again provides them with reassurance. DIY stores like Home Depot in the US and B&Q in the UK try to hire older men as store colleagues. This is based on the idea that young customers, especially women, are more likely to request help from and believe advice from a silver-haired, father-figure type than a typical store helper who may be a teenager. The sense that they will have vast experience provides not just further information about the issue but also further reassurance about the solution. The symbolic hierarchy of expertise is often used as a competitive tool with each competitive brand trying to demonstrate that it is more expert than the other.

When purchasing large-ticket items such as a car, house or pension women experience a certain amount of anxiety. In this situation women would prefer to buy brands that are experienced and have a long history in that area. This provides them with emotional support and reassurance as well as advice. The reduction of consumption anxiety is crucial to converting sales whether they are large-ticket items or not. The recent L'Oréal campaign for shampoo draws on this factor. Its advertising taglines, "Because I'm worth it" and "Because you're worth it," seek to reduce anxiety over the purchase and provide reassurance.

Strong eye contact

Women are better able to read body language, and a high level of eye contact is a classic pointer of explicit trust. People who avoid eye contact either are very shy or are trying to hide something. This is because eye contact is one of the easiest ways to tell whether someone really means what they are saying. Research studies (Baron-Cohen 2004) have proven that even at an early stage, girls pay more attention to other people's faces, whereas boys prefer to look at inanimate objects. The research assessed both newborn babies and young girls and boys to identify whether their attention would be more drawn towards the human face and particularly the eyes. Without knowing the sex of the child or baby, researchers consistently found that young girls spent more of their time gazing at or glancing across to the human face rather than to an inanimate object. Baron-Cohen concludes that this suggests that the differences are biological rather than socially driven. This means that women are born to be more attracted to the human face than are men. This continues throughout their lives as they are more likely to seek and enjoy good eye contact than men. For an advertisement, this means that using images of people with their faces or eyes obscured or out of frame will have a detrimental effect on an advertisements appeal to women. Using people's eyes as a focus of advertising imagery has always been the best way to gain emotional response from advertising imagery as it helps to build a connection between the image and the audience. Often advertisers use images that have people facing away from the viewer as though looking to the future to indicate a positive effect. But it has the opposite effect: semiotically it suggests to the viewer that they are not important enough to be faced or looked at, which creates a negative impression.

Intimate language

The themes of conversation are markedly different between women and men. Women prioritize social themes such as relationships, clothes, hairstyles, food, health and wellbeing. Men tend to talk about things: cars, technology, gadgets or sport. As we have seen earlier, women are better able to articulate and discuss their emotions. The language they use is more emotional, often describing how they feel about what they want rather than the object. This difference is crucial to the tone of voice and language that should be used when communicating with women. Not only is the content more emotional, but so also is the language they use to describe it. Therefore the depth of emotional vibe the listener receives is dramatically higher from women than from men. Because the entire conversation is loaded with emotional cues it is therefore much easier for women to empathize. It is not surprising then that men find this more difficult, leaving their conversations more emotionally barren.

> Word-of-mouth marketing is the number one way to get your message through to other women. Nothing beats a personal recommendation from the sisterhood.

Women are very good at using the disclosure of a secret to strengthen a relationship. The secret may often be a simple one like a confession of a fear or weakness such as a fear of heights or a weakness for chocolate. But the act of self-disclosure demonstrates a strong element of trust in the listener and encourages them to reciprocate in turn. As people begin to build a relationship with someone they slowly release or reveal more intimate information about themselves. This mirrors the level of trust that they believe lies between them. Women offer more intimate secrets about themselves much earlier in the relationship than men, often within minutes of meeting. When two couples meet for the first time, it is usually the women that build a rapport quicker than the men, who will remain at a surface level of conversation. This is driven by men's competitive attitude that giving away too much personal information will leave them more vulnerable and in a weaker position. Women's interest in and ability to discuss secrets is often described as gossiping. The fact that these conversations are frequent and contain large quantities of secret information is the reason that men think that women gossip too much. Anthropologist Robin Dunbar (cited in Baron-Cohen 2004) describes gossiping as the human equivalent of primate grooming. He suggests that gossip is a type of social glue that binds us together and encourages us to develop mutually beneficial relationships. It is easy to see that women, as strong empathizers, like and enjoy more frequent conversation than men. The unrestrained nature of gossip conversations means that there are no taboos or restricted topics. For marketers this means that no one will hold back from sharing their delight or disappointment in their product or service. Men will rarely be overenthusiastic about a commodity unless it is a technology product. They will also be far less critical of a product they have bought because it would reflect badly on their judgement. Women see sharing secrets, even about poor products they have bought, as a way of helping others in the group, while men's first concern will be to avoid personal embarrassment.

HOLISTIC

Women, as the stereotype goes, are much better at multitasking than men. Characteristics noted earlier in the book have highlighted differences in brain structure and ability to connect different parts of the brain. Their greater nurturing and collaborative skills all add up to women's higher ability to juggle several different parts of their lives at the same time. A typical day for a working mother includes not only completing her paid professional day job but keeping the family running, attending to such matters as healthcare, schooling needs, shopping, cooking and cleaning. As all women know, if they were paid per profession that they actually undertake then most women would be very rich indeed. Even when an enlightened man is part of the relationship, women still typically undertake far more of the nitty-gritty tasks around the home. Some women are expert at juggling a wide range of things, but would appreciate brands that keep them going and support them on their continuing journey, or which, if they are feeling unable to cope, help them get back on their feet so they can continue supporting the family and juggling all those responsibilities. A recent painkiller campaign highlights the message "Gets you back on your feet instantly" and shows a woman able to quickly continue her multiple roles as mother and worker. Some women feel exhausted from these multiple roles and are realizing that they can't have it all. Clearly this presents the marketer with two very different mindsets requiring different propositions and messages. Women are better able to combine activities so that they are more efficient at keeping family life progressing. They will perceive great benefits from brands that can be used in parallel with each other. For example, they like banking services that can be phoned while preparing their children's school lunch in the morning before work. Services that are located in close proximity to each other allow women to kill several birds with one stone. They can prepare the evening meal while chatting to friends and organizing their next vacation; all from the kitchen table.

Women's brains are better connected across all areas so they are able to grasp the big picture meaning better than men. This means that focusing on the wider emotional benefit of a product or service is more effective than emphasizing only a few product features.

This multitasking model presents new opportunities for marketers. Marketing messages need to be combinatory and suggest how things fit into a wider picture. By contextualizing the product, women are better able to recognize their own situation and perceive the added value of the brand. Women do not want sympathy about the complexity of their lives. What they want are things that support their ability to perform within that complex context. Messages about how easy it is to manage things, often while doing other tasks, are extremely powerful. Such a thing might be a new cosmetic product that stays looking good longer so that women don't have to keep replenishing it through out the day. Or a banking service that sends an account balance update every morning to their mobile phone, helping them to better manage the family's finances. When a woman is feeling ill a remedy will always be well received if it can be taken easily with no side effects, thus ensuring that she can continue apace through out the day. Marketing messages need to portray women as mastering this complex life rather than being overcome by it. Women recognize that they are very busy, but don't want to be portrayed or perceived as not able to cope. They definitely don't want overly patronizing imagery or tones.

In order to promote health and well being, Wal-Mart sponsors "Speaking of Women's Health," a non-profit group that seeks to empower women to take responsibility for their health and wellness. In addition to in-store displays, the company sponsors seminars at 3000 stores. Each Wal-Mart venue commits to facilitate one in-store event per year that offers female customers health screenings and advice from the store's pharmacists. This type of community-based initiative has helped Wal-Mart be one of the United States's leading socially responsible companies and speaks clearly to women's wider perspective on life.

SUMMARY

This chapter has identified the core attitudes and needs of women in general but this is still a relatively blunt way of describing half the world's population. The next three chapters look at three different segments of women to gain a deeper understanding of the attitudes these women have in addition to the core ones we have just examined. The first of these is Generation Y women, who have a world view that they have a single, whole life, rather than one that has separate subcompartments for work, family, personal, social and private. Generation X women try to manage and balance the two main buckets of their lives: home and work life. Finally, Baby Boomer women have the most traditional lives.

All women share a set of primary characteristics and needs based on the physiological and brain differences discussed earlier in this book. However, women and men are not polar opposites with extremely feminine or masculine traits; they sit on a continuum of femininity and masculinity. Marketers need to identify where on this spectrum their target audience is located in order to properly focus their marketing campaigns.

The first of these primary characteristics is empathy. Women have a profoundly greater ability to see things from the other person's point of view and to subjugate their own feelings for those of others. This is in sharp contrast to men, who remain egocentric.

The next key differentiator is that women have a higher emotional intelligence than men. They focus on the feelings of others rather than the achievement- and goal-centered men. Women use this emotional intelligence to build valuable relationships with others. These relationships are deep and long lasting. They provide a strong-word-of mouth network that marketers need to tap into if they are to be effective.

Women are also more deeply concerned about nurturing others. This social support is a reciprocal arrangement within the sisterhood and ensures their emotional growth.

Women have higher communication abilities than men and this enables many of their other needs and attitudes. Without sophisticated communications it would be difficult for women to create such powerful and emotional relationships.

Women in general have a strong need to nurture their self-esteem. This characteristic is driven partly by their greater sense of place and empathy with others in the group. But is also is moderately driven by the broadly patriarchal society we live in where women are still subjected to inequalities.

The final characteristic is their ability to see the big picture. Women's brains are much better at connecting all the parts of their lives. Women also have to manage far more aspects of daily life than men typically do. They often have to be mother, partner, worker, nanny, and cook all within one day.

3

Researching women's needs

We use this chapter to introduce the best way to define a specific target audience of women for your brand. This begins with female segmentation fundamentals and is followed by detailed examples of how demographic, behavioral and attitudinal segmentation techniques can be applied to different female audiences. It highlights the main female life stages and their implications for marketers. Three of the most important demographic and attitudinal female segments – Generation Y, Generation X and Baby Boomer – are examined in exhaustive detail in the subsequent three chapters of this book. One of the most important aspects of long-term brand management is the ability to manage female customer relationships across the transition points in a woman's life journey. Understanding what women want is just as difficult as understanding what men want, but the research techniques we need in order to find this out are somewhat different. It is crucial for marketers to accept this issue. Because women are different from men, they require different techniques to unlock their attitudes to the brand. It is no good simply avoiding the differences or expecting our normal methods to deliver insightful results if the research methods are not attuned to what we already know about women. This chapter examines the latest methods for identifying uniquely female needs and desires.

OBJECTIVES

- Identify key segmentation parameters including demographic, behavioral and attitudinal segmentation techniques
- Identify key female life stages and their implications for marketers

- Identify the most effective qualitative and quantitative research techniques and tools to understand uniquely feminine needs and desires

WOMEN COME IN ALL SHAPES AND SIZES

Research into female customer groups and needs is a relatively immature science. Its immaturity is confirmed by the rapidly changing landscape of research techniques that fall in and out of favor with organizations. Until very recently a simple demographic targeting was all that was required and assumptions could be made about purchasing habits based only on occupation, age and gender. Equally, just a few years ago a program of focus groups was expected to unlock the hidden secrets and latent desires of customers. Today there is a growing recognition that women do not all fit the same stereotype and that there are just as many professional focus group attendees as bored participants who frankly often lie about their intentions to project their preferred image. In particular there are few research programs specifically designed to illicit women's needs and desires. Most companies do not distinguish between men and women in their research techniques for gender neutral products and services. Given the attitudinal and behavioral differences already identified in this book it is important to develop new techniques that specifically unlock insights about women's purchasing attitudes and usage behaviors. There are many emerging ways to better elicit more detailed insights about women customers.

> Too often women are lumped into an amorphous mass. Great marketing to women comes from detailed insights about a specific attitudinal segment of women.

There is also a realization, more than ever before, that a narrower target audience definition and deeper insights into that audience will massively aid the development of original brands and marketing communications. Many organizations struggle to distinguish between their current user base – the people who actually use their product or service – and their desired target audience. The target audience is a much smaller group of women toward whom they are going to direct their outbound marketing activities. These are the ones that are the ideal customer for a brand. They are also the customers who are most likely to influence the rest of the user base and therefore be a powerful ally in driving sales. The target audience should be a sharply defined attitudinal segment that will be both highly profitable for the brand and one the brand can service to higher standards than the competition. The most important result that any research program can deliver is a powerful insight into why one woman buys your brand over another.

MARKETING SEGMENTATION PARAMETERS

Segmentation of markets lets marketers focus on specific types of women and their needs. Savvy marketers know that clever framing of the segmentation model can add significant value above and beyond the research within each segment. This means a marketer needs to conceive creatively of a different way of looking at the world their product or customer belongs to. This might be to change from thinking about running trains to focusing on connecting people; or switching from thinking about mobile phone technologies to an approach that is based on relationships. A fashion brand inhabits a world that is not about keeping warm but about representing a style of life. A hotel is not a place to sleep but one to relax and revitalize the soul. The more uniquely the world of the customer can be reframed, the more likely that research will uncover a competitive advantage for the brand manager. Failing to reframe the customer's world is also the main reason many brands fail to achieve a truly sustainable advantage. Reframing starts with exploring the boundaries of the segmentation. This might involve refocusing on the purchaser rather than the user of a child's board game like Monopoly. It might be about refocusing on much younger but more desirable future customers, rather than current users. Once this has been confirmed, the type of segmentation will help determine the level of research required and whether qualitative or quantitative evidence is more appropriate.

Demographic

Demographic descriptions and segments are the foundation of gaining insights into women customers. These form the basis of an initial cut of the female market. It could be about focusing on a certain age of women with a magazine like Saga for the over-50-year-olds. It could be that a luxury hotel or spa brand wants to target well-off women that earn over $100,000 annually. Or a vacation brand like Sandals may want to target only couples for their resorts. This type of segmentation is easy to undertake because the information is readily and cheaply available and provides a broad cut of the female population. It is difficult to misuse this information but it also gives very little deep insight because it is a relatively blunt segmentation tool. In order to develop compelling products and services marketers need to gain insights far beyond the summary of typical demographics, namely:

- age
- marital status
- occupation
- education
- social grade
- income

These represent only the most obvious characteristics of a specific group of women.

Behavioral

Further behavioral descriptions about a woman's lifestyle add more detail to a target audience but are still broadly mass-market. Behavioral segmentation means dividing an audience by the actual behavior a group of women exhibit. This can be WHERE they go for or consume your product or service. For example, it can be based on where they prefer to shop – the local high street or the out-of-town mall. Or where they prefer to buy their DVDs – at a specialist entertainment store or online from Amazon.com. Behaviors also include WHAT a woman likes to buy. For example, do they like to buy multi-packs of products to save time and money, or individual ones for personal convenience? Does one type of woman like to buy her food at the supermarket, while others prefer a specialist shop like a delicatessen? The final behavioral segmentation slice is the HOW of women buy things. Some women prefer to buy their food shopping once a week while others prefer to buy it as they need it throughout the week. An airline may want to become the airline of choice with women who travel frequently throughout Europe.

Some of the WHERE, WHAT and HOW of women's purchases might include:

- brands they purchase
- leisure activities
- shopping habits

- product usage patterns
- media watching preferences

Attitudinal/psychographic

The biggest insights into what women want are gained though psycho-graphic or attitudinal profiles that explain the WHY women purchase. These add the necessary richness and texture that allows genuinely new insights to be drawn about what kind of product or service they may prefer. Psychographic descriptions include what motivates women, such as:

- maximizing their time with family
- saving time
- buying the best available
- being a good mother, lover, friend
- concern for the environment

They may also include a summary description of their attitudes to life and are often used by marketers with stereotypical caricatures like "cultural capitalists," "thrill-seekers" and "soccer moms." These customer segments have all the characteristics of tribes. They exhibit shared and sometimes insider values, language and behaviors. As tribes they also tend to display mutual dress, haircut and makeup styles. "Light Greens" are a current segment that display measured environmental concern but only so long as it does not attenuate their lifestyle too much. They will recycle packaging, use low-energy lightbulbs and drive a hybrid car. But they will continue to take long-haul holidays; not use public transport or give up their fresh fruit (flown in from South Africa during the winter). They would like to do the right things like using organic foods and recycle, but at the same time feel rather overly worthy for doing so. This is in sharp contrast to "Dark Greens," who are a segment that will use public transport, actively manage their carbon footprint and be more parsimonious in their consumption habits.

Thinking creatively and insightfully about what kind of segmentation param-eters are best will unlock most of the value of any segmentation study. It's the new way you look at your customer's world that delivers the true business benefit.

The most common pitfalls when developing a segmentation model are:

- Not fully defining the business implications of the segments. For example, size, offer, financial, pricing, contribution or growth.
  - Segmentation needs to be a tightly focused activity to produce good results. Don't try to use one segmentation to solve all your business issues.
- Getting stuck on the jargon; misuse of terminology.
  - Look at what these women are doing and then classify it yourself.
- Using dissimilar characteristics to define the segments.
  - If you don't understand a segmentation that is most likely the reason.
- Confusing current state with future state.
  - Often hides or blurs the gap of what to do to make a woman love your brand.

LIFE STAGE

One of the most effective ways to research and market to women is through insights into their life stage. In modern society it is no longer possible to simply describe women by their age. Life stage segmentation combines the benefits of demographic and attitudinal segmentation to create a rich mosaic of the woman's life. There is a growing trend to extend life experiences and challenge historical conventions. A woman who is pregnant may be 17 or 43 years old. The kind of woman she is may vary greatly, but her overriding current emotional state and needs will vary much less as they are driven by her life stage. Older women may have more experience but the newness of becoming pregnant will reveal new challenges to both these mothers in similar ways. They will both be equally grateful when a brand offers them support by recognizing their situation and providing fast-track channels for them in stores or additional colleagues to help fetch items for them to review rather than having to walk miles around the store. The benefit of taking a life stage view is that marketing solutions can be built around attitudes and daily lives. This ensures that the solution is more relevant to the audience than to any corporate dogma. Life stage segmentation provides richer insights into a woman's life because it focuses on her attitudes and emotional needs rather than just the factual demographics we are told they should be experiencing at a specific age. Even within life stages, there are substages: the beginning of trying to become pregnant or postnatal needs; or when a girl turns into a teenager compared with when that teenager goes off to university or starts her first job. Tesco the supermarket chain is renowned for using a segmentation system that covers around 42 different subsegments. This illustrates the need to avoid over-simplistic generalizations. This book describes many of women's attitudes and behaviors in sharp contrast to that of men's. But as discussed in the introduction, this is simply to heighten the point being made. In the real world the infinite variations of different women need to be better under-stood as groups of individuals that are experiencing similar things at one point in time. This will change as they grow older, but by identifying needs based on a present context we at least ensure high relevancy for those targeted.

Much of the scientific evidence used at the beginning of this book illustrates the combination of biological and social influences that help define women. This journey is obviously more influenced by biology at the beginning, then less so in the middle, while biological influences increase once again as we get older. The journey for women has one monumental difference from men. For women there is a finite time period in which they can become pregnant and reproduce. Whether they would like to be a mother or not, a major part of a woman's life journey is already determined and there is little that science can do about it. Future generations will con-tinue to stretch the boundaries of that time period, but never overcome it. The age between teenager and mid life is biologically best for women to ensure successful reproduction; but it is also the prime years for women to have fun and build their careers. After all, that's exactly what their male friends will be doing during this time. Like so many complexities of life, just when modern women are most likely to want to be free, their bodies are probably telling them they should be reproducing.

> Life stage research is more effective with women because they experience more discrete chapters of different needs and behaviors in their lives than men, although men require the stamina for the long haul and weight of continuous career progression.

Women generally go through a set series of life stages that are far more distinctive than men's. This shapes the world that is their context for any marketing and purchasing decisions. Clearly there are several main events in a typical woman's life including childhood, puberty, marriage, child-birth, childcare and middle age – aside from education and professional life (Table 3.1). Men go through many of these stages as well. But the timetable is more fixed for women and this applies additional pressure around these life stages. In order to connect emotionally with women through market-ing, it is necessary to show that the brand recognizes that these life stages are happening. This does not mean patronizing or overly simplifying these issues. There should be positive acknowledgement and reinforcement that a woman is still an individual throughout. Understanding each context in detail helps outline potential marketing differences. One of the most pro-found things that a brand can do during this period is recognize and support women who may rightly feel that this is unfair.

Young girls

There are many DNA and hormonal differences between women and men but their initial upbringing until the age of three also plays a significant part in their gender characteristics. Otaki *et al.* (1986) undertook a study to assess the differences between three-month-old babies in different coun-tries. Their study of American and Japanese babies mirrored the country differences observed by Hofsteede (1994) in his research. Otaki found that boys in Japan were significantly noisier than girls while American girls were identified as significantly noisier than the boys. These initial dif-ferences are probably due to the behavioral conditioning by their mothers when being noisy is maintained through to American adulthood. We all rec-ognize the stereotype that American women are more talkative and noisy than American men. And Japanese men are dominant and loud compared with the more submissive and quiet Japanese women. These theories about child development highlight the important point that even at an early stage baby girls are remarkably different from their baby boy siblings or friends. Parents can often feel that they have trained their young to admire pink or play with dolls rather than tanks by overenthusiastic present-giving. But the reality is that at this stage of their lives girls have far less testosterone and far more estrogen running through their bodies than their male counter-parts. Some of these biological characteristics direct their social ones. So girls tend to more closely watch and imitate their mothers than do boys their fathers. Women are more likely to have face-to-face contact with babies, exchanging emotional communication and developing emotional bonds. Women are more likely to subjugate their own wishes for those of their

Table 3.1 Women's life stages with accompanying hormonal, brain and reality changes (Source: Brizendine 2007).

Life stage	Major hormone change	Female-specific brain changes	Reality change
Girlhood	Massive estrogen secretion from 6 to 24 months	Verbal and emotional circuits are enhanced	Major interest in playing with other girls not boys
Puberty	Estrogen, progesterone and testosterone increase and begin monthly cycle	Increased sensitivity and growth of stress, verbal, emotion and sex circuits	Major interest in sexual attractiveness, desperate love interests and avoidance of parents
Sexual maturity	Estrogen, progesterone and testoterone change every day of the month	Earlier maturation of decision-making and emotional control circuits	Major interest in own wellbeing and not damaging the fetus; coping with fatigue, nausea and hunger; surviving the workplace
Pregnancy	Huge increases in progesterone and hestrogen	Stress circuits suppressed; brain calmed by progesterone brain shrinks; hormones from fetus take over brain and body	Major interest in wellbeing, and not damaging the fetus; coping with fatigue nausea and hunger surviving the workplace
Child-rearing	Oxytocin, cycling estrogen and testosterone	Increased function of stress, worry and emotional bonding circuits	Major interest in wellbeing, development, education and safety of kids; coping with increased stress and workload
Menopause	Low estrogen and no progesterone	Decline of circuits fueled by hormones	Major interest in improving wellbeing and less interest in taking care of others

child, such as playing the child's favorite game when they want to. In contrast, fathers are more likely to impose their own will on the choice of game or entertainment. These examples suggest that young girls are more likely to mirror their mother's nurturing, attitudes and behaviors, which drive choice of clothes, foods and more feminine colors. Marketers can learn a lot about what young girls want by looking at what their mothers want. While this should not preclude treating young girls as individuals, using their mother's attitudinal preferences is an excellent proxy.

Stereotypical evidence about girl's higher academic achievements is also true and provides clear direction for marketers. Girls have more advanced communications skills, better language, listening and a greater vocabulary than boys do, at an earlier age. This advantage continues throughout women's lives. Any marketing communications should take advantage of this fact even early on in life. Advertising campaigns can use a higher eloquence when speaking to young or teenage girls than they can speaking to boys. The boys, who use a more intuitive process of reading only headline-grabbing messages to make decisions, would find it difficult or unattractive to be communicated with in the same way as the girls.

Teenage girls

As girls grow up to become teenagers they transition through puberty. While boys do this as well, girls tend to start and finish the process earlier. Some girls will start this earlier than the norm and this tends to have an adverse effect on them. By contrast, boys who start this process earlier tend to benefit from this by being more relaxed, popular and athletic. This is because the rite of passage to becoming a man affirms a boy's virility in a way that increases his leadership and status qualities. Girls find the process more distressful and consequently early-maturing girls are more likely to be emotionally confused than their male counterparts. Marketing to teenagers is always a difficult task. They are developing their own sense of identity, separate from that which is governed by their parents. Brands can help girls by providing them with role models and identity cues that are inclusive. By encouraging girls to feel part of the group they enable girls to both use and receive nurturing emotions. Martin Lindstrom's book *Brand Child* (2003) examines these issues in more detail.

Puberty is a time of great dichotomies for girls and boys. These range from matters of sexual identity and orientation to religious and spiritual beliefs; from health and wellbeing to hedonism and sobriety and entertainment and sport among others. Each one of these areas offers a variety of choices that teenage girls must make to help solidify their sense of self-identity. With these choices come enormous freedom and opportunity. For brand managers this fragmentation of identity represents a major shift from a relatively narrow group of needs and expectations to an explosion of divergent needs and desires. Brands need to distinguish their values in order to consciously attract teenage girls otherwise the very nature of their broad appeal will be highly offputting to many subsegments of the market. The teenage market is as always the most difficult to appeal to because they are extremely cynical and change their point of view frequently. Brands that can support teenage girls in making these choices will build lifelong relationships with them. As they go through this cathartic period any brand that provides them with direction and substantiation of their own values will be seen as a lifeline within the complex teenage years. Teenagers undergo so many changes that brands need to define which parts of their lives they would like to belong to and which they are unable to support effectively.

Generation Y women – career starter

Once beyond the watershed of the teenage years, women study as well as work within a society that is still patriarchal. This is despite the fact that in the UK and US almost half of the working population is women; 45 percent of women work in the UK (EOC 1996). Inequality over pay still exists, with conservative estimates suggesting that women are paid between 75 percent and 85 percent of their male colleagues' salaries (Burr 1998). The attitudinal shift from meek stay-at-home girl to confident businesswoman is happening swiftly: girl power is here to stay. This change presents women with another dichotomy; in order to get ahead in a (still) man's world should they emphasize their feminine wiles or should they revert to masculine stereotypes of the "bitchy female" in order to succeed in a masculine, competitive and aggressive environment? Thankfully, the move towards service economies and emphasis on customer relationships favors women with their superior group-oriented, nurturing, empathy and listening skills. Brands need to deal with women in a way that respects their femininity rather than stereotyping them as either hardline or overly soft. Marketing campaigns need to address women in ways that acknowledge their intrinsic abilities as women and the benefits of these. The zeitgeist is moving in their favor and brands that can exhibit an early understanding of this will quickly gain first-mover advantage. Once enough women believe that a brand has their true interests at heart, the female grapevine will soon guarantee brand success. It takes a bold marketing director or CEO to make this change explicit in their organization; but as we have seen in the introduction, the potential rewards are huge.

Women in their early to mid working years recognize the underlying biological time frame and will be looking to accelerate any opportunity for brand and life experiences. While a man may take a more linear approach to engaging in life experiences, women need to try it all and try it now. They need to quickly sample things so that they can decide if they are worth further investment. Brands that give women an opportunity for mini taster sessions will receive a huge welcome from women. These could be sampler product sizes that help women to trial new ideas or brands. They could be experiences packaged or bundled with standard products they already buy. It could be a cereal brand that offers free cereal snack bars for a trial period. Or it could be a shampoo sampler for a new brand extension that is bi-packed with their normal brand. Men often wonder why women are so keen on these; it is because they provide ideal first-hand research into the product before making the final purchase decision.

Family manager

The most distinctly different life stage for women is obviously childbirth. This does not exclude men, but we need to recognize that the physiological impact and requirements are different for women. The period spent during pregnancy has huge hormonal effects on women and often distorts self-image. Feelings of unattractiveness are not uncommon, despite the fact that this time should be about celebrating the wonderful state of motherhood. Following childbirth itself, postnatal depression is again not uncommon and encouraging a sense of purpose and wellbeing is vital. The first few months and years will see a woman adding yet another major role to her life as the mother and nurturer of the child. This doesn't mean fathers don't contribute, but even today the reality is that women still often do all or most of the housework. The main message for women is that they are coping with it, and brands that take a supportive tone are more likely to be appreciated. Again this needs to be a subtle message; obvious patronization or condescension will be derided or ignored. One of the most significant social groups following childbirth is the young mothers' club. These are highly influential and word-of-mouth marketing among their members will be incredibly effective. Young mothers have so many new things to learn at this stage that advice and guidance help reduce stress and make them feel they are coping well. Genuine collaboration between brands and mothers at a grassroots level will create loyal shoppers for a lifetime. Marketers need to play the long game with women in this life stage. Take the time to really understand their needs and provide them with objective and appropriate products and services that make women's lives easier.

Little emperor syndrome

Everyone recognizes that children have an emotional effect on our purchasing behavior and women, owing to proximity, bear the brunt of most of this. This is particularly pronounced with first babies. The temptation for young mothers (and fathers) to buy the best for the little one is all-powerful. After all, who would take a chance with the most important thing in their life? This results in the "little emperor syndrome" that is much evident in places like Asia and Italy. Children are doted on and rather spoilt well into their late teens and beyond. Again, marketing that is objective, impartial and responsible is most likely to be welcomed by young mothers. It is important to resist the temptation to over-emotionalize and dramatize the issues, as women will be turned off by such cynical approaches.

Generation X women – working mother

One major difference between women and men is that often women's careers are a series of broken periods, unlike men's which tend to be continuous. Women are better able to cope with this, since they are more adaptable and better at multitasking. However, one of the emotional hurdles they must overcome is having the confidence to return to work after childbirth. Again the marketing message for women at this life stage is that they are able to cope and continue and are brilliant in having achieved so much. Any reinforcement of their ability and supportive messaging will help encourage women purchasers. Women also realign their priorities following childbirth. They may have been keen on experimenting with lots of new things but post-childbirth their feminine attitude turns towards nurturing. This means that they are more willing to displace their own needs for those of their child. Men, by contrast, still prioritize their own needs and ambitions far beyond that of their children. Brands need to dramatically change their stance from the expansion of opportunities for adventurous women to one of trusted commodities within the comfort zone of now more cautious women. This shift is difficult to achieve without a portfolio of brands. The outcome of this change is that women may well stop trying a large number of brands and become more "mono brand" in their decision-making. Brands that have already secured the trust of young mothers will reap the reward of both loyalty of sales and ample opportunities to cross-sell. Modern working conditions are making life easier for everyone to take career breaks and work part time and with flexible hours, these changes especially aid women in their ability to continue juggling many tasks at once. The title "Superwoman" has been coined for high-profile women like Nicola Horlick, the recent superstar asset-manager, wife and mother of several children. The reality is that every working mother is a superwoman, every day of her life, and she knows it. She doesn't need the label – what she needs is to be supported in this role.

Baby Boomer generation – maturer women

As women are getting biologically older, they are also getting psychologically younger. Women who are now fifty feel like forty or younger. Rising health has annulled the need to be physically cautious, and this has resulted in an explosion of desire and will to regain an expansive, pre-childbirth lifestyle. This presents several specific opportunities for marketers. As women endeavor to recapture their youth, they expect to be communicated with in a youthful manner that belies their years. Inside they may feel like a 22-year-old; the last thing they want is an overly serious brand telling them what to do. In particular, the Baby Boomer generation are having a ball. It is no wonder that they are one of the most important targets. They typically have ample savings, something that younger generations of women have not even begun to do. They have paid off their mortgages – something that younger women may never do, thanks to rising houses prices. They are experienced and have seen changes in the world that they want to explore. They have the ideal combination of wealth, health and opportunity to do exactly what they want. Marketers need to make these women feel that they are in control of their own destiny. For many of these women this will be a new sensation, as they grew up in postwar austerity conditions and in a time when a Protestant work ethic was admired. Brands need to be suggestive with these women. They cannot tell them what to do but neither should a brand be completely passive; it needs to encourage these women to learn how to go beyond their own expectations. Any brand that can convey this sense of liberation will find a willing audience. Like any supportive and educational relationship, these women customers will reward a brand with clear loyalty for having collaborated on their renaissance.

Statistically women live longer than men – approximately five years longer in most Western countries. They also remain mentally and physically active to an older age than men. This means that there are many products and services that will shift from being targeted at mature couples to focusing on widows. This may seem harsh but it is a reality that we see everyday. Lively groups of mature women outlast and outshine their slumbering male partners. Brands that target Baby Boomer women need to be highly sensitive to their potential new single-woman status. They have spent a lifetime in partnership, sharing the roles and responsibilities with their husbands or partners. They have to relearn many skills that may have atrophied over time. Brands need to be highly sensitive to the changes these women face. They are probably deeply traumatized by the loss of their partner, yet are required to swiftly develop a wider set of skills. Their perspective will also shift from the joint to the singular and potentially open their horizons. Brands that can transition or straddle this divide can play two roles. They can either be supportive throughout this period, acting as a solid foundation during troubled times, or they can shift their emphasis by increasing the volume on certain aspects of their brand personality. They might move from a "well-trusted family friend" positioning to one that supports the opportunity of the sisterhood. Sub-branding or portfolio products and services are the most effective way of managing this change in audience.

Transition points

There has long been an academic discussion about whether women go through a continuous journey of change in their lives or a transition between separate chapters of a story. Men largely move along a continuous journey, albeit at different speeds and intensity throughout their lives. The evidence seems to be that women transit across a series of distinct phases in their lives. This means that brands must be prepared to either lose women along that journey or adapt themselves via sub-brands or changes in marketing approaches in order to retain a woman's custom. Neither of these is especially easy and requires a high degree of integrity if these strategies are to succeed. The starting point is for marketers to identify these transition points in their target audience. Again these must be life-stage-driven rather than purely age-driven. The next step is to define the values that will be retained across the transition point and those that need to be discarded or adapted.

The following chapters describe in detail three of the largest and wealthiest segments of women: Generation Y, Generation X and Baby Boomers. Generation Y women are those born after 1978, who are optimistic, confident, image-driven, gregarious and fluid, in control of their work–life balance. Generation X women were born between 1966 and 1977. These women are somewhat cynical and feel a strong responsibility to the world. They are more individual and want reality and quality rather than fantasy or the superficial. They try to structure an effective work–life-balanced existence. Finally the Baby Boomer women are those born between 1946 and 1965. These women are young at heart. They want to be included in the latest things and do not want to be patronized as old. They are the first generation that is healthy and wealthy enough to truly make the most of their advancing years.

EFFECTIVE MARKETING RESEARCH TECHNIQUES

There are many qualitative and quantitative research techniques that can be used to gain deeper insights into women. This book focuses on describing those that have proved highly effective in recent years rather than being a book about general research techniques. The following techniques are mostly qualitative and align particularly well with the biological and behavioral differences specific to women. Women are much better than men at masking their true needs because they are more likely to put the success of the group above their individual needs. Researching men is therefore relatively easy as they are more transparent and direct about what they want. Researching women's needs, because they are more hidden or latent, requires subtler and less direct forms of research inquiry. Ethnographic research methods help marketers to uncover latent needs and allow women to reveal their desires without group bias. There are several types of ethnographic research that are all especially suitable for researching women.

Ethnographic methods

The world's most valuable brands and greatest new products are usually born out of a clear insight into the target woman's life. One of the easiest ways to gain these insights is to take part in their lives with them, observing how and why they do things in their own home. This can be done on a macro or micro scale. It may mean living with a family and observing their breakfast rituals to understand why they buy both butter and margarine (The mother liked butter but thinks the margarine is healthier for her children.) In order to research a girl's soft drink it may require hanging out with teenage girls, in the playground or park. This requires observing their attitudes to each other in groups or the language they use to describe their friends food and drink. It is also effective in building up a picture of the brand world of the target teenager, by observing what they wear and which products they use. The key to great observational research is to not only observe the "what" people are doing but to understand the "why." What motivates women to behave this way and what emotion does the activity create – either negative or positive? The "why" is the starting point for any brand or product development; it is the outcome the woman is trying to achieve. It is crucial to start with the desired outcome and then reshape everything to deliver that outcome with the minimum effort.

Diary-keeping (video and verbal)

A diary of daily events can reveal insights into women's priorities in life and into the specific product or service. Because diaries are simple and convenient, they encourage the women to edit their day down to the key events, issues and feelings. Getting women to keep their own diary ensures that their observations are not tainted by the prejudice of the observer. The women should be encouraged to illustrate their diaries with doodles and other examples of what they were feeling throughout the day. These diaries enable the marketer to understand what role the new brand or product plays in their life. It is worth noting that women tend to be much more diligent than men at completing a diary of this kind.

Participation

Recently, a consultant working for an entertainment brand was required to research this market through participation. This meant she had to stay at every major casino in Las Vegas to understand the complete competitive context. She was required to watch every show, play every game and analyze the total experience, atmosphere and customer engagement. By actually taking part, she was able to participate in the experience first hand within the authentic context (unlike a focus group which is in a neutral room). She experienced the full range of emotions for herself and acutely observed others as well. This also enabled the researcher to observe the target audience and gain deep insights into their wider emotional and brand world.

A, E, I, O, U ANALYSIS

This is a form of ethnographic research that explores the context of a brand and the customer relationship within this context. The design strategist Nick Durrant helped create this approach and it always creates unique insights for clients using it. For this type of research marketers need to identify the current state, but also suggest the ideal future state within these contextual boundaries.

A is for activities

This element is informed by how women think about and categorize activities as they happen. Marketers need to assess these from both the user's and the brand owner's point of view as there may be a huge dissonance between the two. Using the customer journey provides a clear structure for this analysis. It is important to see what the presumptions have been made about when these activities occur or are enabled and then liberate them from hackneyed marketplace presumptions: "I'm a global nomad, not a family resident in Birmingham, and may need to do things in an order different from what others expect . . . " Capturing these insights means focusing on something emotionally descriptive about each activity; the bathing experience is dreamy, the eating experience is gourmet. For a mid-market hotel brand the key list of activities might be: booking, checking in, dining, sleeping, gym, breakfast and checking out. Thus for example:

- *Old dining experience.* This is based around prescribed "proper" times and day-parts – for example morning breakfast, afternoon tea, evening dinner, nighttime sleep.
- *New dining experience.* A simple daylight/nighttime division to manage but without a prescriptive interpretation – that is: "My stomach says it's breakfast time even though it's four o'clock here . . . "

E is for environments

This element is informed by the semiotics of site placement, architecture and interior design. There is an increasingly sophisticated dialogue between global cues and local cues that need a singular vision and point of view to be successful. For a mid-market hotel this could involve using a guest designer such as Terence Conran, or Ian Schrager's favorite designer Philip Starck, to envision and execute with discipline based around an attitudinal concept, rather than just decor; for example:

- *Old environmental experience.* "Global hotel with local accent."
- *New environmental experience.* "Local hotel with global smarts amplifying "local placeness."

I is for interactions

This element is informed by the semiotics of service design, staff behaviors and uniforms. There is a spectrum of interaction design from the highly prescribed to the highly intuitive. This is covered in more detail later in the book. In the hotel example this might be for example:

- *Old interaction experience.* Heritage job roles with many handoffs – for example concierge, doorman, check-in staff, porter.
- *New interaction experience.* Modern flexi service with single named point of contact to look after my interests – "Hello, I'm Jim. Let me sign you in and take these things off your hands. Is there anything you need to do immediately that we can help you with?"

O is for objects

This includes everything from appliances to fixtures and fittings. Like environments, these need to be curated from a central idea in order to clearly communicate their message. Without this ruthless editing, the range of objects that appear in most brand experiences is a strange collection of legacy, corporate-brand-approved and personal items that leaves customers bewildered; for example:

- *Old object experience.* Lowest-common-denominator, generic functional design semiotics – "Here's a clock-radio and an iron and a plasma screen."
- *New object experience.* Sensitively selected, emotionally appealing objects designed with a point of view, "ease of use" and a personality – "That's a fun bottle-opener."

U is for users

This element works best when derived from an attitudinal profile rather than simple demographics. It is also important to identify the coping strategies people use to overcome less than ideal experiences. We need to be vigilant to constantly assert what we really mean by "enjoys authentic experiences," or "has sophisticated tastes" – sophisticated for whom and in what context and to what degree?

- *Old user experience.* Backward-looking – "Confirm my expectations."
- *New user experience.* Future-looking, authentic experimenter – "Surprise me."

Combined AEIOU experience

We can draw together all five elements to create overall insights into both the current and ideal future experience for the fictitious hotel example.

- *Old.* "Classic but generic" with "conspicuously discrete," heritage luxury – "tasteful."
- *New.* "Energetic and personal" with "seamless continuity," new luxury today – "tasty."

Marketers can often learn far more from things women don't like or don't do than from things they like or enjoy. This is called provocation research. Women often find it easier to identify and recall things that they didn't like more than the positive aspects of a brand. This is because negatives cause disruption and a high emotional charge in our daily experiences and therefore are more likely to be remembered. There are several techniques to help us gain deeper insights through the use of examining the negative aspects of a brand or market.

- Qualitative research provides many insights on women. The key is to be bold and selective enough to choose only one of them in order to build a strong, polarizing brand.

Deprivation research can help to assess the influence your brand has on a woman's daily life by taking it away. By depriving someone of their usual brand, marketers can gain insights into two things, first, what they miss most about not having their favored brand and second, the needs and attributes that another brand satisfies better than their usual one. Often it is easier for women to articulate what they don't like about a brand than what they do. Deprivation research uncovers many of these negative dissatisfiers as well as satisfiers.

> Conflict triumvirates are groups of three people with divergent attitudes to a particular product or issue.

These groups facilitate greater tensions between each other's point of view. This means that each person of the triumvirate has to explain to the others why they like and dislike components of a brand. The prescribed tension encourages greater clarity and ensures strong engagement with the research respondents. This works better with women, who by their nature often try to avoid conflicts in a traditional focus group. Men tend to have no problem with competitive conflict in focus groups. It also puts the target audience or current buyer under pressure as it means they must themselves reassess why they like and dislike certain brands. They may change their attitudes when they hear the other side of the argument or they may reconfirm their commitment to a brand. Either way the process of articulating their loyalty reconfirms a brand's strongest positioning or message.

A recent research program for a premium wine brand that wanted to enter a new market used this technique to uncover powerful insights. Instead of just gathering a group of target women who all drink premium Australian wine and attempting to unearth new insights, the client used conflict research to help shape their market entry program. This involved using three different types of women in each one-hour session:

- a young mother who mostly drinks wine at home
- a lawyer who heavily drinks beer with work colleagues in a city bar
- a charity worker who drinks regularly but not heavily with friends at the local pub

The results were successful in identifying the premium wine brand that had a "cheeky" side to it that could be highly attractive to the target women customers.

Semiotic research and analysis

There is a more detailed explanation of semiotics later in the book; but it is just as useful in researching brand associations with women as it is in helping to define the brand in the first place. Semiotics is the study of signs and signifiers that people use to interpret meaning from the things around them. It is particularly useful in defining the role and relationship women have with language and messages. Semiotic research is a specialized skill that requires experience to accurately identify the appropriate interpretation. Semiotics is especially relevant to researching with women because they possess much greater sophistication in speech and listening. Semiotics relies on subtle definitions, and the greater the linguistic ability the more insightful the results.

QUANTITATIVE RESEARCH

There are a huge range of quantitative research techniques and this book will confine itself to the brand driver analysis that can unlock facts about the branded component of purchase decisions by women. By its very nature, quantitative research is more applicable to researching both women's and men's needs with fewer anomalies than qualitative research. The influence of brand on purchase decision is different for different categories. For largely functional categories such as utilities, the brand influence is small. For more emotional purchases such as perfume and cosmetics, branding plays a more influential part based on the perception of the brand and the accompanying smell (see Figure 3.1).

The influence of the brand within each customer demand driver denotes the level to which branding plays a part in each criterion. Some criteria will be barely influenced by brand, while others will be mostly influenced by the brand. Price for example is in itself not influenced by brand, although the perception of price can be. Equally, for a gasoline station, location is one of the key demand drivers, but this is barely influenced by brand. Women have a huge variety of functional and emotional needs. In order to make sense of these it is important to categorize them as demand drivers that relate to specific purchase decisions. These will highlight what a woman's purchasing criteria will be and how much weight each one has in her purchase decision. Functional demand drivers have a fixed satisfaction driving ceiling, whereas emotional drivers have an exponential ability to drive satisfaction (see Figure 3.2).

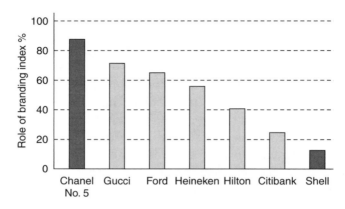

Figure 3.1 Examples of the 'Role of Brand' and its influence on purchase decisions.

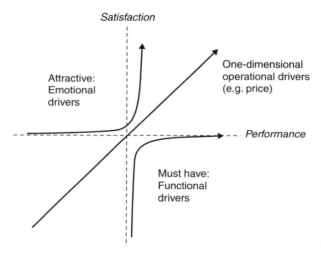

Figure 3.2 Emotional drivers demonstrate a continually increasing impact on customer satisfaction, unlike functional drivers whose influence plateaus.

It is important to understand the role that brand plays both in your category and within individual purchase drivers of demand. This will help drive optimal marketing investment across these values and touchpoints.

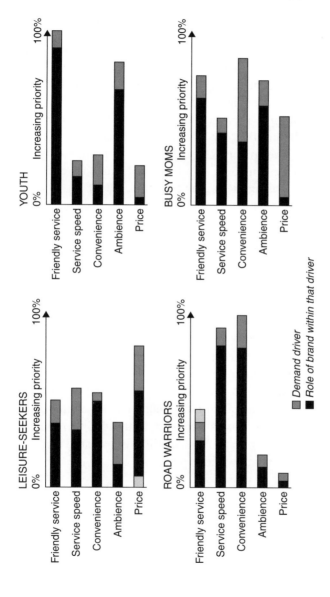

Figure 3.3 Demand drivers and the role of brand are prioritized differently across four different customer segments.

This role of brand analysis can be used to calculate which parts of the customer purchase decision are most heavily influenced by the brand's attributes rather than functional drivers of choice (see Figure 3.3). For an airline, the timetable and destination network are largely fixed functional drivers of choice for customers, who either want to fly to that destination at that time or not, while the decision about the level and style of service are chosen based more on the brand attributes exhibited by each airline and therefore the role that brand plays in these decision drivers is far higher. It is important to recognize that not all customer demand drivers can strongly influence brand-building. This level of prioritization allows marketers to focus on demand drivers that can be highly leveraged in order to ensure a greater return on investment (ROI). The role of brand analysis, which requires a data-driven calculation, should be performed for each individual segment to further focus marketing investments.

SUMMARY

This chapter has identified the most effective approaches to segmenting female audiences through demographic, behavioral and attitudinal segmentation techniques.

Women in particular pass through a series of life stages that reflect their distinct biological makeup. Women's life stages are more distinct than men's and require careful marketing strategies and tactics to traverse the transition points between them.

The chapter concluded with detailed descriptions and benefits of the key differences in qualitative and quantitative methods and techniques required to specifically research women's needs and attitudes to brands. Ethnographic, semiotic and other non-formal qualitative research techniques can provide stronger insights into women's needs.

4

Generation Y women

PROFILE

Generation Y women exhibit specific secondary characteristics in addition to the primary characteristics all women possess. Generation Y women are optimistic and believe life is for living. They are confident and have high self-esteem. These women are able to move fluidly across the traditional boundaries of work–life balance and in their ability to buy both premium and value brands with the same level of pride. They are in control of their lives and are able to play the game of life to their advantage.

Women born after 1978: UK and USA (see Figure 4.1)

- UK: 4 million women (Carat 2005) = 13.3 percent of the female population (total female population 30 million; www.cia.gov).
- US: 36.2 million women (Johnson and Learned 2004) = 24 percent of the female population (total female population 151 million; www.cia.gov).

Key events that shaped their world

- genetic engineering
- fast internet communications
- virtual life
- ethnic diversity
- 9/11 and 7/7 acts of terrorism

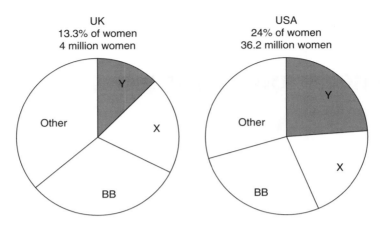

Figure 4.1   Generation Y women in the UK and the USA.

INSIGHTS

Generation Y women are aged between 18 and 29 and come after Generation X (aged 30 to 44). They have more in common with their mothers. A recent research study by the *Washington Times* revealed that 30 percent of Generation Y women want to have children in their mid twenties, while a further 50 percent want them before they are 30 years old. The majority of Generation X women (60 percent) by comparison preferred to wait until they were at least 27 years old before having their first child (Widhalm 2006). Like their mothers, the Baby Boomers, Generation Y women believe that the ideal age for marriage is 24 to 26 years old, while only a third of Generation X women believed this. Generation Y women are more interested in what you are like and like doing than defining you by what you do for a living.

Generation Y is at the bow-wave of a rapidly growing world with exploding statistics, like 300 million Chinese people under the age of 24, that's 147 million young Chinese women (Timeasia.com 2006). They are the first generation that is entirely free from ties with China's troubled social past. They have not been trapped like their parents in the "iron rice bowl" system of social provision and control, under which the state provided everything and there was no choice. The parents of Chinese Generation Y women ate, slept, worked, married and grew old at their work "unit." With everything provided it meant that they had little concept of choice or consumerism.

Chinese Generation Y women, however, have had to acquire basic life skills as they seek their own apartments, choice of food, clothing, work and partners. In the largest cities these women are "well dressed, constantly talking on their mobile phones and jam the city's Western style bars. They are ambitious and confident and work for Western firms and internet start ups" (Time Asia 2000). These young Chinese women are described as "Generation Yellow," the first independent group of Chinese; while their parents, who lived under the strict Communist regime, are described as "Generation Red." Generation Yellow women are heavy mobile phone users and the *China Daily News* (2006) recently reported that total mobile phone usage would reach 520 million in 2007, or one third of the entire population.

India has an even younger population with 50 percent under 25. That's over 510 million young people, 252 million of them women (Paul 2005). This group of young Indian women is roughly the same size as the entire population of the US and demonstrates the dramatic changes within these emerging powerhouse countries.

These women are extremely independent, having grown up surrounded by social fragmentation, job insecurity and parental instability. They come from an era where their parents are likely to have been divorced. But growing up with a single parent was more commonplace for these women and less socially stigmatized than Generation X women. They are also well loved and optimistic about the future (NAS 2006). They are more confident of living life on their terms; while Generation X women feel controlled by events, Generation Y women feel in control of those same life events. They have a far more optimistic outlook than Generation X women, again more similar to the Baby Boomer generation.

> This generation of women believe in a combined whole life and do not separate their career, fun, relationships and responsibilities into separate buckets.

Generation Y women recognize that a job for life is unrealistic and that the employee now has the power to pick and choose the conditions under which they work. They value flexibility and diversity and are therefore more likely to be portfolio career women. They would rather run their own business around their lifestyle than climb the corporate ladder (Fragiacomo 2006). They see and want to be part of the big picture. They are pluralists at heart and are comfortable with ambiguity in any part of their lives, whether at work or in their personal or family lives. They are independent, confident and comfortable with taking risks. Generation Y women are more likely to become entrepreneurs than return to the corporate workforce which they see as inflexible to their personal needs and situation (Widhalm 2006). This means they are more intuitive about the decisions they make and can be opportunistic rather than organized and structured in their thinking. These women are the most financially independent, with the average female American college graduate earning annually almost the same as their male counterparts for the first time in history: $24,000 to $26,000 (Francese 2003). That means that these women no longer have to negotiate with their partners over purchases and have the credit to fund their own independent lifestyle. It also means that when these women do marry, they and their spouse's combined income will be significantly higher than that of the Generation X and Baby Boomer couples at the same point in their lives.

Equally Generation Y women are less interested in a fixed, long-term plan and prefer to live for the moment. They want instant gratification, believe more in short-termism and tend to do things at the last moment. They try and achieve what they want with as little effort as possible; they are not the grafters of the world. They are ambitious but only on their terms; they are not prepared to work all hours and sacrifice their lifestyle to achieve greatness. They are excellent at building social and work networks and use these to facilitate their lifestyle. They would rather spend quality time with people than achieve results; for them it is the journey rather than the result that counts. They are as anti-establishment as the Generation X women but they do believe in the power of a cause and frequently are fervent supporters of causes and rights campaigns. They are however less interested in committing to or taking responsibility for traditional obligations and in this sense tend to be rather maverick and difficult to pin down. These women grew up in the most ethnically diverse community ever and expect that this is the norm anywhere they are; it is no longer a question whether diversity is part of social and work fabric, but a fact. Work, life balance is a key tenet of Generation Y women. They are not interested in older, authoritarian models of power and management. They are not slaves to the organization, but look for a firm that will support their ideals, which might be environmental, workers' rights or just plain work–life balance: a firm that recognizes them as an individual rather than a mere corporate resource.

They are heavy users of technology although not in the geeky way that men use it. They are well connected and use Instant Messenger to connect with their friends and family. They are the first entirely computerized generation and their education was based on computer learning. This affects not just their appreciation of technology but their conception of how the world works. They expect retail and service brands like Gap, Topshop, Zara and Banana Republic to have detailed computer records of their purchase history, even when shopping at their non-neighborhood store. They expect that they will receive statements and bills from brands like Verizon, Orange, Bank of America, ING direct or First Direct via email or mobile phone rather than primarily by mail. They gather their news and views online, at time.com or theonion.com rather than buy a newspaper, because it's real time and up to the minute. It's also already summarized and they can search for favorite topics easily. They share information via gmail.com or myspace.com and youtube.com, while downloading the latest tracks and clips for their iPod or laptop. Generation Y women are optimistic and idealistic about the future in a way that reminiscent of the Baby Boomers and consequently are often called "echo Baby Boomers" in research circles.

MARKETING STRATEGIES AND TACTICS

Optimistic and idealistic

Generation Y women (see Figure 4.2) have a very positive view of the future and a self-belief that they can achieve what they want in all areas of their lives. They want to associate themselves with brands that do not just show the future through futuristic visions, but are clearly leading the way to that future. Brands like ebay.com and myspace.com succinctly reflect their desire to be well connected but offer an alternative way of doing business/ shopping/connecting with their friends and interests. Brands should create online media that offer a dialogue with these women rather than the traditional push media approach. The recent RED campaign is a great example of a brand that appeals to Generation Y. These women like it because it has a strong ethical cause and gives them a chance to engage with the brand in a two-way dialogue. These appeal to their sense of commitment to a cause and the active empowerment that it gives them as individuals. Using RED products like the Amex RED card confers on these women identification with a wider world cause way beyond just a credit card. Even surprising brands like Shell and BP can be appealing to Generation Y women because they at least are attempting to devise a future scenario that encompasses both the need for transportation and environmental restraint. These

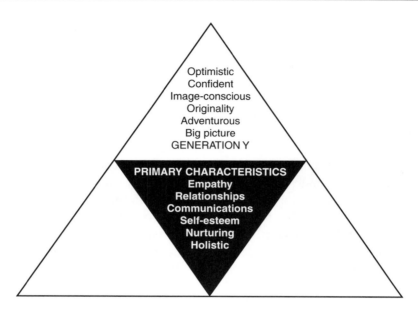

Figure 4.2 Primary and secondary characteristics of Generation Y women.

brands appeal because they are actively attempting to write the future rather than passively allowing it to happen. Disney is another brand that strongly appeals to Generation Y women by communicating a sense of idealistic belonging. You don't just visit Disneyland, you join its family. It is an all-encompassing world that is pure escapism for everyone. The brand is built on the values of family and on providing magical entertainment. This is just the kind of fantasy that Generation Y women enjoy as it is both active and highly optimistic. Its present and future is always positive, avoiding any sense of reality. This is in contrast to Generation X women who would find it too false and cynically dismiss it as utopian or kitsch.

The mobile phone company Orange, with its strapline "The Future Is Bright," has always been the archetypal Generation Y brand that expresses their values of human optimism and emotional appeal. This contrasts starkly with the Vodafone and Verizon brands which have presented a masculine brand image to the market. Both of the latter express their brand through a series of technologically futuristic images. Semiotic analysis of those images suggests an austere future with little warmth, a picture much more of individuality than sociability. Its messaging focuses on performance characteristics rather than community value. T-Mobile portrays a Teutonic version of the world, where everything is precise and rigid. Orange has used its optimism to show that it believes communication can make the world a better place. It is a brand that has a cause at its heart, an activist in the mobile arena. Orange pioneered "pay per second" billing rather than per minute, to show their customers they wanted a long-term relationship and not just short-term profits. Orange recognized that if their future was to be bright, then they must make their customers' lives individually brighter first. Not to be outdone, China Mobile recently launched the "M-Zone" brand that targets Generation Y consumers (m-zone.com.cn). The brand has a youthful style with fluid lines and uses cartoons of animals in sneakers to generate streetwise coolness. Its funky graphics using art-style imagery of exploding drinks bottles help to enhance the anti-establishment stance of the brand. It's exactly the kind of image that Chinese Generation Y consumers want to project. These women want to distance themselves from large corporations like the parent brand China Mobile. The China Mobile brand was only launched in 2000 and it is already the highest-ranking brand in the *BusinessWeek* Best Chinese Brands 2006 league table where the value of the brand alone is £18.6 billion. This is not surprising given that the mobile phone market in China is 520 million users, up by 60 million users (the entire population of the UK) in the last year alone. That means that there are 5 million new users signing up each month.

Confident

Generation Y women have higher self-esteem than either Generation X women, who are often described as the "Lost Generation," or Baby Boomers who suffered from a stronger paternalistic culture when they were growing up. They are neither angst-ridden nor downtrodden and are keen to make the most of their lives. Recent research in Asia has highlighted that Generation Y women are becoming more independent than their traditional subservient stereotype might suggest (Lawson and Brahma 2006). Singaporean women are leading this change from the conventional patriarchy, but India and Thai women also express the desire for autonomy and independence. In comparison, Chinese women express much less of a desire for individuality and their society as a whole is still based on social responsibility and respect for the family unit. The root of this new-found liberty is attaining financial freedom from men so that they can make their own choices in life. These changes are at the heart of the social changes transforming these countries. There is a shift forward from the subsistence lifestyle where women and men were simply working towards a basic lifestyle. Women in particular have often suffered because it was typically the men who were the only wage-earners, making women entirely reliant on their partners. As these societies have become wealthier, the women's focus has shifted from the group and family towards themselves as individuals. For Generation Y women this has coincided with other cultural changes like increased materialism, easy access to mass media, entrepreneurialism and the relaxing of religious conventions. All of this has altered Generation Y women's perspective on their role in society and encouraged them to think about themselves more in the ways typified by the following insights from a Thai woman: "I'd like to be prepared, to keep my options open; society doesn't pressurize single women like it used to." An expert text on India concurs: "Women can be confident on their own today; now you see young women asking themselves: what is it that I can do beyond being married" (Lawson and Brahma 2006)

> These women are independent and confident enough about their future and are eager to take control and lead others.

One of the key areas that have changed in the past few years is the amount of effort and money spent by these women on their appearance. In a recent research study in India, hair colorants are now mostly bought in the metropolitan areas for reasons of fashion, rather than covering up gray hair (IMRB 2007). In most Asian countries there is still a large divide between consumer activities in the urban metropolitan areas like Singapore, Mumbai, Shanghai and Seoul and the rural hinterlands. However, these metropolitan areas are driving consumer desire and aspirations even within the rural neighborhoods.

For Generation Y women, their heroes are other women – and not the ones that are merely beautiful but those that have true substance too. They have confidence through their ability rather than just their looks. These icons include those listed by *Forbes Magazine* as the most powerful women in the world: Angela Merkel, the German chancellor; Condoleezza Rice, the US secretary of state; Wu YI, the Chinese vice premier; Han Myung-Sook, the new South Korean Prime Minister; Wu Xiaoling, deputy governor of the People's Bank of China and India's Lalita Gupte and Kalpana Morparia, joint managing directors of the huge ICICI bank (forbes.com 2006). The list goes on, but the central thread is that all these women have achieved greatness through ability at the highest level. They are the inspiration for Generation Y women who gain confidence from their talent at breaking through traditional masculine barriers in work, politics and society. They no longer want to be told what to do and marketers need to avoid patronizing these women by telling them which brand to choose. While women in general have a greater sophistication in their communications, Generation Y women are likely to prefer a more direct approach. They do not expect or need to use convoluted or oblique language. They favor directness because they have greater self-confidence and also appreciate brevity.

Image conscious

Generation Y women have a clear understanding of how the brands they purchase affect their complete self-image. They are acutely aware that what they purchase, where they work and where they socialize all partly define who they are as a person. They are the most media-savvy of generations and use and abuse brands to their own ends. They need to know that a brand's environmental credentials are impeccable before they will become an advocate. Making personal statements through their consumption of brands is crucial part of their lifestyle. Brands clearly need to pay attention to Generation Y women and identify the key themes that will appeal to them. Generation Y women are keen to be streetwise, do not suffer the cynicism and negativity of the Generation X women and do not take themselves too seriously. Motorola began to target Singaporean Generation Y women with a new campaign. The company had recognized that modern women use their mobiles not just as a communications tool but as an expression of their lifestyle and individuality. They focused on repositioning Motorola Singapore as a lifestyle and entertainment brand rather than a technology firm (Yin 2006). This meant creating marketing activities that spoke directly to these women through channels that they had not previously expected. These activities kicked off with the Motorola Super-StyleMix '06, a nine-day event of fashion catwalk shows, art installations and parties. Other events included co-branding with *Singapore Women's Weekly* and launching the MotoMusicShow on a popular Singapore radio program. The combined effect helped change perceptions of Motorola in the minds of these fashion-savvy Singaporean Generation Y women. It's a great example of a brand that is not afraid to tailor its offer to the specific needs of a group of women.

The Apple iPod is the perfect example of a product that blurs traditional boundaries and strongly appeals to Generation Y women. It is a consumer electronics product that doesn't look masculine and aggressive. It is remarkably simple to use and underemphasizes its huge technical capabilities. Just like Generation Y women, it has created its own new market rather than being an extension of a previous one. The product design follows Apple's brand ethos of humanizing technology. Their brand seeks to build the technology around human needs rather than hoping people will adapt to technologically focused products. The video recorder, a typical Baby Boomer product, has always been perceived as a paragon of poor interaction design. Its design is focused on the proliferation of technological features, subsumed behind a façade of masculine styling. The way to use it has been totally ignored at the expense of the way it looks, resulting in a triumph of styling over usability. Even the remote control with its plethora of minute buttons shouts its technological credentials far louder than its user-friendliness. This kind of product design has been overtaken by a more user-centered approach of which IDEO, the Californian design company, are pioneers. Their most celebrated design was for the first computer mouse; imagine how different life would be without those. IDEO together with Apple have changed customer expectations about the levels of product usability and styling. They are at the forefront of the feminization of product design, abandoning superficiality, aggressive styling and over-complication and replacing them with the emotional appeal and intuitive usability that perfectly suits Generation Y women. Apple's groundbreaking iMac started the trend by reinterpreting the humble computer as a lively, colorful object rather than a "slave gray box." This was swiftly followed by the sleek iPod portable music player that changed our relationship with music, from a short collection on a tape, record or CD to a jukebox full of downloaded individual tracks. The shift from matte black, the color of the electronics high priesthood, to shiny white marked a further shift to demonstrate the lower aggression levels of the new line of products. The iPod with its shuffle track feature is a terrific expression of what Generation Y women want – flexible access to everything – and they are happy to break the convention of listening to music by specific album and genre. It is the ideal metaphor for Generation Y.

Generation Y women love to shop for cosmetics at Sephora. While the store has debranded its shelves in favor of color coding, it is still one of the chicest brand to use. The concept behind the store is that there are two graphic zones: the black zone is for perfumes and the white for cosmetics, while the rest is a huge red carpet to give customers that million-dollar celebrity feeling. The store has been successfully rolled out around the world and J C Penney the US department store is opening Sephora concessions within a thousand of its locations as a way of attracting younger Generation Y women to its stores (BrandNoise 2006).

Originality

The boom in repertoire shopping, where customers are happy to mix and match their shopping purchases – what they wear and eat for example – across a wide range of luxury, mid range and value stores has been driven largely by Generation Y women and their ability to transcend traditional boundaries. For retailers this means these women may buy cut-price commodity items like toilet rolls, washing powder and food basics at the discount center like Lidl, Aldi or Wal-Mart, while buying luxury items like special cheeses, hams and deserts at a delicatessen like Fresh and Wild or Planet Organic. Supermarkets have responded by creating luxury own-label offerings like Tesco's "Finest" or Wal-Mart's "Organics" offerings. Both these propositions encourage women to mix and match the quality of their foods from basic to luxury but within the same store. These women do not feel the social stigma or guilt attached with buying basic grocery brands that Generation X or Baby Boomer women might. Their self-confidence allows them to be proud of these choices and feel more like a savvy shopper in doing so. This is a fundamental distinction in the way Generation Y women think about themselves and the brands that they buy. They see themselves as in control of the situation and mastering the options rather than being dominated by a brand's status.

While the Generation X woman are self-conscious and cynical about marketing and branding in general, Generation Y women are able to use it to their own ends. They enjoy the overt control they have over brands and this is demonstrated by their ability to combine value brands like Target in the US or Zara in Europe with high-fashion brands like Gucci, Prada or Burberry. Their effervescent confidence encourages their self-expression far more than previous generations. This drives their ability to break the fashion rules that might constrain the more conservative Generation X women. Repertoire shopping is also a result of further fragmentation of consumer segments. This means that marketers need to be highly targeted when they define their female target audience. In these chapters we have examined in detail the three most financially attractive female subsegments;

but even these require further attitudinal descriptions to be viable for a specific product or service brand. This is especially true for younger female segments like teenagers and twentysomethings, all of them particularly difficult to target with marketing communications as they tend to be highly cynical audiences. However, marketing with a repertoire range of sub-brands and own-label brands helps to keep these segments loyal and happy.

> These women have a clear sense of their own style and they love brands with strong personalities.

Zara, the Spanish fashion brand, has risen rapidly up the *BusinessWeek/* Interbrand (2006) best global brands league table to number 73. Its brand now has a value of $4.2 billion as its exports itself further across Europe. This is in sharp contrast to the fortune of the American Gap clothing brand, once a generation favorite, which has tumbled in recent years, losing 22 percent of its brand value in the past year alone. The reason for this is that the Zara brand helps Generation Y women project a distinctive, optimistic and self-confident image that is as much at home in Madrid as it is in Manchester or Munich.

To build brand awareness among Generation Y while preserving its underground image, Red Bull utilizes originality to gain cut-through. Red Bull remains "under the radar" while driving activation with the following tactics to increase appeal to Generation Y:

- *Student brand managers.* Red Bull provides student reps free cases of energy drinks, asking them to organize a party. The "brand evangelists" spread positive word of mouth quickly and inexpensively and offer credibility to a product in an increasingly crowded beverage market.
- *Consumer education.* Red Bull utilizes shiny silver cars with large reproductions of a Red Bull can be attached to the back. The drivers distribute free Red Bull to interested consumers, building a cutting-edge image cost-effectively.
- *"Extreme" sponsorships.* Red Bull supports communities of extreme athletes to maintain its "liquid adrenaline" image.
- *Concentrated product.* Red Bull offers a single size and flavor of product with no brand extensions. To emphasize the product uniqueness, packaging consists of 8.3-ounce cans, which counter traditional 12-ounce soda cans.

Many car brands have recognized that Generation Y women want their car to express their personality and the relaunched Mini and VW Beetle and the Mercedes Smart car and are perfect examples of brands that achieve this. These women are not shy about proclaiming their personality and their cars need to stand out from the crowd. The usual ranges of Toyotas and Peugeots simply do not have enough character for these women. They do not accept the distinction between a "work" car and a "personal" car because they are not influenced by this division in their minds. While Generation X women try to constantly manage and balance these two requirements, Generation Y women have risen above this and are less bounded by such established divisions.

In fall 2002 Saturn launched the Ion, a small car targeted toward Generation Y at a very reasonable price. Saturn, already a very popular brand among the target, utilized low-pressure sales techniques and fixed prices to drive customer satisfaction among Generation Y-ers. In addition, Saturn sponsored the Goo Goo Dolls' summer tour, and displayed the Ion as a co-sponsor at college football games at the Summer 2002 X games in Philadelphia, a skateboarding and bicycling competition. It successfully connected with these women in a non-threatening way.

Virgin and Richard Branson are both classic Generation Y brands despite being significantly older than this generation. Branson sees enjoyment of life in everything, work is about merrymaking and his airline is a party in the sky. His confidence, zest for life and sociability all resonate strongly with Generation Y women. The Virgin brand fits clearly with the Generation Y vision that life should be fun first and functional second. Branson's enthusiasm for life and ability to make work seem like it's a fun game synchronize neatly with Generation Y women. He leads by example and his informal style contrasts strongly with that of other global corporations. Employees are encouraged to share their ideas for improving the business with him directly. When he travels, he always carries a notebook and chats with the aircrew to find out how they would improve things. These conversations may be about their uniforms being uncomfortable or better ways to serve their customers. He sees the two parts as indivisible; in order to keep his customers happy he has to keep his employees happy first. Virgin has a reputation for having great employee benefits like the annual party for the entire company or the range of incentive programs. He also embodies a maverick spirit that shows that the little person (although not so little now) can take on the big firms and win by focusing on what consumers want. There is an energy and excitement to the Virgin brand that is the antithesis of the mundane or bureaucratic. Virgin demonstrates a powerful Generation Y tribal element that helps it outperform other businesses. It treats passengers as friends right from the moment you book with them to when you are flying. The company tries to make the experience as entertaining as possible. In its friendly and collaborative approach Virgin is the opposite of British Airways, which has a strong masculine brand ethos. BA also delivers great service but in a style very different from Virgin's. Flying with Virgin is about joining the party; flying with British Airways is about corporate leadership and superiority. Generation Y women are out to get the most out of life and do not see why they should compromise on this enjoyment. They are also not bound by the conventional hierarchical structures.

Similarly, easyJet in the UK, the US jetBlue and SouthWest airlines have redefined the travel experience of low cost flights. These low cost airlines are now some of the most highly rated in the world with top customer loyalty and advocacy scores (Gahilan 2003). They have created this success through a compelling service experience that not only makes employees and customers feel good, but they are also more likely to recommend jetBlue to their friends. Their focus on being friends with their customers emphasizes the human and interpersonal nature of their ethos. Their core values of safety, caring, integrity, fun and passion indicate that the experience is always going to be more about enjoyment than hard work. Generation Y women do not suffer from the social stigma that Generation X women feel about travelling with low-cost airlines; their confidence and strong self-esteem means that they see it as the savvy choice and they are happy popping over to Lyons or Bologna for a weekend for a bargain price or down to Austin, Texas for a similar treat.

Adventurous

Generation Y women are less conservative than their Generation X elders. They are happy to take more risks in the pursuit of fun and enjoyment. They have much higher levels of vivacityand joie de vivre and recognize that this is no dress rehearsal. Brands can tap into this attitude with a call to action that encourages them to try out new experiences. Or they can persuade them to go further than they have previously gone, whether it is extreme skiing, deep-sea scuba diving or trying a new nightclub. Nike recently ran a new campaign across Europe, the Middle East and Africa to expand the notion of athletics to encompass women's modern dance. It was designed to challenge the idea that women dancers are not physically powerful or not to be considered as athletes. The television advertisements showed supremely fit dancers talking about how they don't get the same adoration that sports people get in an Olympic arena, yet train just as hard if not harder to perfect their technique. The Nike dance routine supported by a Chemical Brothers soundtrack emphasized the physical and powerful nature of modern dance to demonstrate its legitimacy as an athletic sport.

Online brands that have been traditionally male-focused have been successful at attracting Generation Y women by offering them an emotionally charged experience that mirrors the "girls' big night out." The betfairpoker.com site has launched an online micro site for women called Betfair Poker Ladies' Night. While it is stereotypically pink in its design, it offers Generation Y women a specific entry-point into the world of male-dominated gambling. The tone of the site is designed to tap into the customer's hedonistic attitude and provide her with a fun night in. The site mirrors the main website with easy start options, free roll (initial free bet) for the first 350 women to join the game each Tuesday evening, and upwards to entry to the world women's online poker tournament in Las Vegas.

Games console brands are also targeting Generation Y women with instructional workout videos and immersive game environments Women currently play 75 percent of online puzzles at MSN games and it's a natural step for them to get involved more with highly engaging game formats (Jana 2005). As male markets become more saturated it's an easy gain for manufacturers like Sony with their PS3, Nintendo with Wii and Microsoft with Xbox to focus on becoming more attractive to Generation Y women. These women have the ideal balance of confidence and risk-taking coupled with their desire for an entertaining work–life balance. Developers already sell 37 percent of their video and computer games to women but these are often bought as presents for their partners. They are now targeting Generation Y women with specific titles that focus on their interest areas. Exergames, a hybrid of the entertainment and workout genres, helps these women to satisfy their needs and open up a whole new market. Sony's EyeToy is a great example of a product that has been tailored to women's needs. It provides easy access to exergaming and a low financial threshold by using the standard PS platform that is already in many homes anyway. However, unlike traditional PS games it does not use a standard control but a motion sensor that allows women to enact their workout program by following the onscreen avatars. Importantly, the designers recognized that these women do not want the over-glamorized Lara Croft kind of illustration that are endemic in men's games, but prefer realistic, healthy and trim women as their role model. The players can even scan in images of themselves via the EyeToy camera accessory. It's not all hard work, though; as the user progresses through each of the levels of fitness they can add in their own features like a sound track as a reward. This nascent form of entertainment has proved very popular with women in the US, UK and Australia, with Home Media Research suggesting this is the ideal way to introduce women to console games (Jana 2005).

Generation Y women are have-a-go heroes and will seek out new products and experiences.

Faced with an aging customer base and fierce competition from Honda, Harley-Davidson recognized the need to attract a new generation of riders. Although Harley recognized that small-displacement motorcycles sold primarily to Generation Y comprised the fastest-growing segment in the US, the company did not manufacture this type of bike. Moreover, its brand image – rebellious, classic, and steeped in tradition – did not resonate with Generation Y women. Consequently, in addition to its existing entry-level motorcycle line, Harley introduced a new model, the V-Rod. Harley also launched its Rider's Edge Program, a motorcycling safety course, targeting new consumers, and attracted Generation Y customers through its grass-roots Harley Owners Group, which has 660,000 members worldwide. The company acquired a significant number of new customers, almost half of them women.

Starbucks is another archetype Generation Y brand. It is not a work-OR-life dichotomy that Generation X women perceive but a blend of all facets of Generation Y women's life. The Starbucks brand ethos focuses not on the process and purity of the coffee (like some of the Italian coffee houses) but on the community spirit of the experience. Starbucks is a brand experience that exudes feminine values. This contrasts loudly with the strident machismo of Italian-styled coffee chains that scream masculinity and caffeine-enhanced performance. Starbucks has managed to produce an extraordinary brand experience that is both reviving and relaxing at the same time. Howard Schultz, chairman of Starbucks, illustrates why this is the case:

A product is made in a factory and can be copied; a brand is something unique bought by a customer: a branded experience is a deep, long-term relationship that triggers emotions beyond the purchase. It is timeless and therefore priceless.

The concept of a casual, nostalgic atmosphere combined with friendly staff creates a place that you can stay for hours. This sense of community and emotional bonding is palpable in a Starbucks coffee house. They have deliberately encouraged people to use their cafés as the "third space," the alternative place away from home and the office. This social function is an important differentiator with other coffee houses that make you feel like you shouldn't stay too long. Stepping into a Starbucks is like stepping into a friend's house for a coffee; there is a strong residential sense of belonging. Their interior design evidences several psychologically feminine cues. The use of softer materials and colors (in comparison with the competition) helps to give a stronger sense of cocooning. Adrian Furnham, professor in psychology at University College, London, has often referred to the psychology of the workspace as indicative of the gender of the organization. Starbucks has emphasized its brand femininity by the use of sofas, soft music and old-fashioned graphics. The use of light wood colors throughout helps to make the place feel warm and welcoming. There are no hard granite, metallic surfaces or cold colors in Starbucks – they want you to feel at home, because they don't want to sell you one cup of coffee, they want to be your friend. A quick examination of the foyer of any major corporation will identify a feminine or masculine business. The feminine ones are usually more colorful and use softer shapes and materials. They will have fewer barriers between the employees and the customers and have places that encourage interaction. These businesses have less hierarchical structures and are more open for everyone to contribute to the success of the company. Furnham even points to the kind of food and drink served at meetings. Those that serve tea in a teapot with soft cakes or homemade biscuits demonstrate more feminine attitudes than those that simply use a coffee machine. It is often the small cues that help customers to decide whether a brand fits with their gendered view of the world.

See the big picture

Generation Y women have a more holistic view and see the big picture more than men. They know how everything fits together in their work life and their personal life. Westpac, the Australian bank, has responded to this need by targeting Generation Y recruits for a job rotation scheme that lasts six months in each of four different departments. This gives these women the opportunity to experience the wider firm and gain a better understanding of how the entire firm works together. Marketers need to recognize that Generation Y women will fully research a brand's proposition before making a purchase decision. They will want to balance the expected price–benefit proposition with additional ethical and corporate social responsibilities (CSR) activities of the brand. Only when the total package is well balanced will they become loyal to a brand. They will spend the time to research the details of brands like Fairtrade, a Toyota Prius car or organic food ranges from Asda, Wal-Mart or Tesco. As more and more companies increase their CSR activities, it can be difficult to differentiate from the competition. These women are not looking at each of these elements as a core differentiator, however, but as the standard they expect from a modern brand that wants to do business with them.

> Generation Y women are open to a wide ranges of views but synthesize their own view of the world without the social angst of the Generation X women.

BP is a service brand that has updated its brand and service ethos for the Y Generation. It has shifted its focus to a more world-friendly and inclusive message. BP has demonstrated its commitment to solar and other alternative energy sources as a proof point of its commitment to the environment. The former British Petroleum was an old-fashioned, imperialistic business with a dark-green shield as its corporate identity. It exuded the masculine values of strength, performance and coldness. In 2001, BP recognized a mood shift in the consumer and decided to evolve its brand and service offer to mirror that change. Out went the shield and old values. In came a new brand, based on a stylized sun symbol, or for some observers a bright-green flower-head. The old-style corporation had feminized its image to become more appealing to Generation Y consumers and women in particular who had become more socially aware. One of BP's four brand values is "Green," demonstrating the importance that the company places on shifting its brand and service offer to become more socially responsible. They have invested hundreds of millions of dollars developing solar and hydrogen technologies that harness the world's resources in a more efficient way. They have also tried to change their forecourt offer with sub-brands like "Wild Bean Café." This type of sub-brand is closely aligned with the new BP Helios (Greek for "Sun") brand and would never have worked with the old masculine shield identity. Wild Bean Café is a light-hearted and social way to sell coffee. Its cartoon-style graphics are warm and engaging. They have used this sub-brand to generate a sense of local place at each site to encourage people to linger while they drink their coffee rather than rushing back to their cars. This would probably be viewed as cynical marketing by Generation X women but Generation Y women appreciate the intricacy of the situation and BP's attempts at dealing with this complexity. They believe in the personal freedom of driving their car but also want to care about the environment. The two are not mutually exclusive as they are in the mind of Generation X consumers but just a part of their multifaceted, contemporary life. Compared with their competitors, the other major oil companies, BP has gone a long way from a greedy oil industry reputation. This type of repositioning is likely to have a high impact on all women drivers who empathize with a brand that represents their attitudes more closely.

Mediated lives

This group use the internet and motion pictures as primary channels for their news and information. They are less likely to read newspapers (Projectbritain 2005). Motion pictures are particularly appealing to Generation Y women with their combination of glamour, rich experience and celebrity. Mobile communications and interactive advertising will increasingly play a priority role in these women's lives. Generation Y women were the first generation to use computers and the internet as a key learning tool at school. According to a recent study, 83 percent of UK Generation Y group use the internet regularly (Armstrong 2006). Women in this segment are highly influenced by brands and internet brands in particular because they are developing their sense of self-identity. These women are using the web for a variety of different tasks. First they are staying touch with friends and family with email and instant messenger (IM) programs like MSN messenger. According to AOL, a new user joins the IM community every 3.5 seconds and its own AOL instant messenger (AIM) boasts over 100 million users worldwide (aol.com). One of the attractions of this form of communication is its rapid conversational protocol that fulfils the instant gratification needs of Generation Y women. Second, these women are using the internet as a series of online magazines like handbag.com or popbitch.com that provide them with lifestyle information on their health, entertainment, relationships, fashion, travel or just gossip. Finally, they are probably the largest segment that is regularly shopping online. Their friends provide them with more support than their family; they are avid magazine readers and their favorite brands include Nivea, Diesel and Apple (Carat 2005). Some of the world's hottest brands like eBay target Generation Y consumers. eBay's brand value has risen 18 percent in the last year alone and is ranked at number 47 in the *BusinessWeek* Best Global Brands league table with a value of $6.7 billion. Its open access marketplace approach to buying and selling fits perfectly with Generation Y women who want to search and buy anything from anywhere. They are tech-savvy and enjoy the inherent global style that eBay's advertising and online experience brings to the chore of shopping. They do not have the cynicism or security fears of Generation X or Baby Boomers and are completely comfortable in this world. They are getting the most from life because they are in control of it and can manage everything around their way of life. Their heroes are Sacha Baron Cohen, Angelina Jolie and Kate Moss, who are all modern, savvy and interesting people who play the game of life well.

SUMMARY

Generation Y women hold all the primary characteristics of women in general but additionally over-index on the attributes of optimism, confidence, originality, adventurousness, image-consciousness and seeing the big picture.

These women have a more fluid approach to life that allows them to manage porous boundaries between work, personal and all other facets of their lives. This fluidity is used by Generation Y women to take control of their lives and get more of what they want out of life.

Generation Y women have a much shorter-term view of life than Generation X women and have a live-for-the-moment attitude that requires more instant self-gratification. They are more interested in making the most of today than in building longer-term foundations for tomorrow.

Generation Y women grew up in a heavily mediated world and are expert at projecting and managing their self-image. They are comfortable with and sophisticated at using brands as part of their image management. This means that brands need to have a distinctive image with a high degree of polarization in order for these women to find them attractive. The more universal a brand's image, the less appealing it will be to these women.

5

Generation X women

PROFILE

Generation X women exhibit specific secondary characteristics in addition to the primary characteristics all women possess. Generation X women are independent and more conventional in their attitudes. They try to balance traditional and modern lifestyles. They have a strong sense of responsibility and are often described as part of the "Lost Generation." These women are more serious than Generation Y women and marketers need to emphasize reality to these women rather than utopian futures.

Women born between 1966 and 1977: UK and USA (see Figure 5.1)

- 20 percent of UK female population (total UK female population 30 million. www.cia.gov) = 6 million women (Carat 2005).
- Just over 20 percent of the US female population (total US female population 151 million. www.cia.gov) = 30.8 million women (Johnson and Learned 2004).

Key events that shaped their world

- Aids
- collapse of communism; fall of the Berlin Wall
- big-bang electronic banking
- rising middle class
- cheap air travel

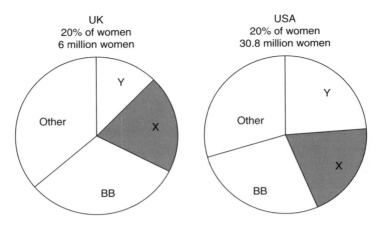

Figure 5.1   Generation X women in the UK and the USA.

INSIGHTS

Generation X women grew up in a time of economic uncertainty. The boom years of the sixties and seventies had left a financial hangover on society and the Generation X-ers were left to pick up the pieces. The sense of job security and final pension schemes that their parents enjoyed were no longer available. They were also too old to take full advantage of the economic boom brought about by ubiquitous internet and mobile phone usage. Generation X women are sandwiched between two more optimistic eras and they feel disgruntled by this. They feel that they have not had the same opportunities as their parents or Generation Y women. They will however have to pick up the pension requirements for their parents as they continue to live ever longer, applying further pressure to their stretched finances. In the US, multigenerational households now account for 4 percent of all housing. While this may not seem a lot, this figure grew by 38 percent in the ten years to 2006 alone (Brandnoise 2006).

Generation X women witnessed the dissolution of many social and work structures. They became the first generation to be worse off than their parents. This was epitomized by the miners' strikes in the UK; the breaking of union power under Thatcherism, aligned with the Ronald Reagan years in the US. Reagan's capitalist ethic drove up unemployment and America's lack of a structured social security system meant hardship for many Americans. In Poland, the Solidarity movement attacked the communist apparatus

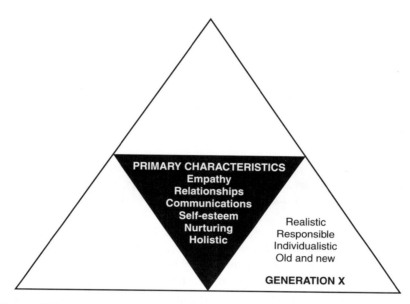

Figure 5.2 Primary and secondary characteristics of Generation X women.

and this was the beginning of the end for Communism as a power block of countries. Japan and Germany were the shining lights in the world economy as the Generation X women were growing up. Japan was the only shining light in the world economy as the Generation X women were growing up. Japan in particular was seen as an unstoppable force, driving cheaper consumer electronics into every home. These economic changes have influenced the outlook of Generation X women, leading them to be less optimistic and more cautious about their lives and futures. Generation X women were born into a world of increasingly high divorce rates and this has made them more independent and self-reliant than the general population. It has also influenced them to seek the permanent structures they may have lacked in their formative years, perhaps through work, their home life or their partners. Generation X women are more likely to be of different ethnic or race origins than previous generations. Their parents experienced the last truly monoculture society. This means that even subsegmenting Generation X may be a blunt definition and may fail to truly represent the nuances of their social or neighborhood networks (see Figure 5.2).

This audience is media-savvy and form many of their opinions from television and radio. They are less likely to use newspapers as a valued informational source (Projectbritain 2005). These women are the first generation to prioritize multimedia information sources. They grew up with the world in color, unlike their Baby Boomer parents who grew up in a black and white world. They are also the first generation of women to face the reality of political and social change brought directly into their homes. They witnessed the horror of the first Gulf War broadcast live as it happened in their front rooms. This reinforced their realistic and non-idealistic views and beliefs about the world.

Generation X women are caught in the middle of the change from the old formal, structured world and the new flexible and open one.

Generation X women are the segment that is currently the best-educated, probably married and with children. They have a significantly higher educational level than their parents, the Baby Boomers. Many are the first generation of their family that received a university degree or postgraduate qualification. According to 2005 US census data, 10 percent more Generation X women attained a college or batchelor's degree than Generation X men. This means that they have a greater emotional and educational freedom than previous generations. They also broke the mould of following their parents' linear career options. University opened their eyes to the wider social and vocational possibilities and gave them a broader world view that informed their life decisions. Generation X women have less faith in the employment system as lifetime employment became a thing of the past. This lack of faith led many Generation X women to focus on knowledge and skills as a form of security that could be transportable and applied to a wider variety of occupations. This also meant that they were attracted to jobs that encouraged and emphasized on-the-job skills training and personal development. It also gave them a cynicism about world issues that their parents lacked.

Generation X women are currently 30–40 years old and at a point in their lives where they are re-evaluating what brings them happiness and how they view work in relation to the rest of their lives. Brands need to demonstrate that they not only understand these changing needs but can also facilitate and fulfil these new needs. It requires brands to be agile or else risk losing their current Generation X customers. One way they can achieve this is to present a range of options to Generation X women. This ensures that they are able to change from their habitual purchase, but still remain with their preferred brand. Clever brand portfolio management will help businesses manage these customer transitions. Mercedes has achieved this with its range of small, family-sized and sporty cars. Women who have spent years driving their family around in a sizable station wagon are now driving around town in a nippy Mercedes SLK. It is the ideal sports car, powerful yet not too unwieldy.

MARKETING STRATEGIES AND TACTICS

Realistic

Dove, the beauty brand, has spent a huge amount of resources redefining its world view of beauty. They recognized that many women found traditional marketing and advertising portrayals as unrealistic and fantastical, for example airbrushed images that showed models who had had their eyes enlarged, their skin smoothed and their bodies slimmed after the photo shoot. While these images might be aspirational they were too far from reality for the cynical Generation X women who want to be treated as adults and not suckered by imagery. This precisely reflects what Dove consistently discovered on its research across the world: "These women found current images of beauty as over-idealized and unattainable." Dove's "Campaign for Real Beauty" is the first time a major beauty brand has spoken the language of Generation X women. Their global research program demonstrates their authentic commitment to reality rather than an airbrushed utopia (Dove 2006). Many of the women shown in their advertising are just as they are in real life; one has a tattoo while another is heavily freckled. The core message is: "All skin is beautiful when it is beautifully moisturized." This is exactly the kind of marketing that will appeal to Generation X women because is authentically real and yet is still sexy because it revalues these everyday women. The financial results are good so far: a 3.4 percent rise in sales year on year – a significant amount in the highly competitive, fast-moving consumer goods marketplace.

> Generation X women are skeptical about marketing and expect down-to-earth realism from their brands.

Philips, the consumer electrical and electronics manufacturer, has recently repositioned itself, shifting from a masculine research and technology company to one that focuses on quality of life. Their new advertising high-lights the emotional benefits of their products. Their new global advertising tagline is "Sense and Simplicity" and conveys their attitude of emotional and sensorial beauty above the traditional, performance-led culture of tech-nology products. Philips and Apple have both paved the way to a more human and less aggressive domestic landscape. They have feminized their consumer's relationship with electronic products and added an emotional dimension to our relationship with these. Under the expert design direction of Stefano Marzano, Philips has also propagated a more radical design approach, bringing real humanity and warmth back into daily product performance. These offer a genuine experience that is not dressed in techno-cratic clothing. Philips's range of products demonstrates the archetype for modern design, marking a new epoch based on quality of life and humanity. Their televisions, shavers and kettles all have a little personality to them, evoking a smile when we touch and use them. They satisfy the consumer with a powerful combination of welcoming styling and minimum features (they just do their job really well).

Volkswagen used its television advertisements in the US to celebrate real life by featuring drivers and passengers in real-life situations. The TV spots often highlight a growing sense of maturity within the drivers, who reflect on situations, such as having a new child, in immature ways, and end with simple, comic taglines that tie in to the company's overall theme, "Drivers Wanted." Volkswagen's strategy was to have a lot of lifestyle ads, which have received an amazing customer response, particularly among women. As a result, Volkswagen ranks as one of the highest car sellers to Generation X.

Take over men's territory

While Generation X women may be as cynical as their male counterparts, they are also blurring the boundaries between the two genders. Generation X women are breaking away from the traditional roles that their moth-ers played and taking on formally masculine tasks. This evidences their strength and independence by competing directly against their male coun-terparts. Men may have become "metrosexual" beings in touch with their female side but women are also taking on traditional male activities like DIY. These women are interested in demonstrating their independence and improving their homes. In the US a new brand, Tomboy Tools Inc., is taking advantage of and expanding this opportunity. Founded in Denver in 2000, it has put down strong roots in the US and Canada and is on its way over to the UK. The basic concept is educational sessions on DIY techniques and tools combined with a female-friendly get-together over a cup of coffee. It is the Tupperware party for the new century. Given the rising number of single women in most Western markets, brands can no longer expect there to be a man about the house to fix things when they go wrong. Several of the new home improvement shows are led by women like Sarah Beeny in the UK, Anna Maurice (the "House Doctor") and Linda Barker. While Martha Stewart in the US leans towards homemaking rather than complete home renovation, they all encourage women to take control of home design and decoration.

Jiffy Lube wanted to make the automotive care experience more comfortable and approachable for women and to generate increased market share among female customers. They redesigned their waiting centers to reflect a new elevated look by installing softer paint colors, higher-end carpet, and comfortable chairs. In addition, Jiffy Lube provided amenities such as women's magazines, Starbucks coffee, satellite television, and web access. With online access in waiting centers, female customers can visit jiffylube.com's "women's channel," which offers tips on prolonging the life of a car, maintaining safety on the road and planning road trips. Based on extremely positive customer response, Jiffy Lube extended the concepts across their fleet.

Generation X women are more skeptical than any other group of women. This means that marketing needs to be very real, and they prefer anti-marketing approaches that avoid hype. Brands need to be genuine and transparent with any claims they make. All women are great at researching potential products and brands and will find out if a brand over-claims or is lees than truthful about its product or service. Featuring an honest straight-forward message – after a decline in sales, resulting from a stale image – Coca-Cola unveiled a $70 million advertising campaign for its lemon–lime offering, Sprite. The ads spoofed celebrity endorsements and other beverage advertising clichés, ending with the tagline, "Image is nothing. Thirst is everything. Obey your thirst." The repositioned soft drink tripled sales in four years. An honest, straightforward message drove success with female consumers.

Using factual information is one of the best ways to convince Generation X women. Factual endorsements help to persuade Generation X women that performance claims are both true and significant. Supermarkets cleverly use endorsements to provide additional provenance to their claims of healthy foods. These might be endorsing marks by partner brands like the Soil Association to create a differentiating message. These go some way to provide Generation X women with an objective third-party approval for the quality of the food. Other brands use product comparison features on their websites to help suspicious customers like Generation X women to check the facts themselves. Financial sites like moneysupermarket.com provide both easy access to the full range of loans, mortgages and insurance products and, crucially, complete transparency on the best deal as well. Similarly, travel websites like travelocity.com and expedia.com allow consumers to track real-time changes in air fares and choose their ideal combination of price and schedule.

Responsible

Brands based on an employee partnership model such as the John Lewis Partnership or the various cooperative societies speak unmistakably to Generation X women because they act as a responsible "family" unit. The John Lewis Partnership demonstrates the full range of responsible brand attitudes, from highly ethical work policies through great employee care and benefits to true working partnerships with suppliers. Consumers find this increasingly appealing and choose these brands over others because of it. Women in particular are strong advocates of such businesses because they recognize and appreciate those mirrored female brand attributes. The retail brand John Lewis uses the strapline "Never Knowingly Undersold." While this may seem a little Dickensian in style, its endeavor is diametrically opposed to glitzy and over-selling claims of other brands. This is a highly appealing antidote for Generation X women who view many brands with suspicion.

The Waitrose supermarket brand answers the Generation X woman's combined need to both make it real and build trust at the same time. Their television advertisements are rich in texture and tone but present real images of real people unfettered by catchy jingles and stylistic effects. The message is that these are people like you and they want to build a relationship with their customers based on trust. Current campaigns emphasize the "partnership" structure that Waitrose and John Lewis both belong to. This is a company that is owned and run by the employees and presents just the kind of explicit honesty that Generation X women need if they are to be confident they are not being cynically targeted by brands. Waitrose is aggressively expanding its relatively small retail footprint by opening 25 new stores in the next 3 years and its plans appear to be financially paying off: like-for-like sales for the supermarket were up 14 percent in 2005 (waitrose.com).

> Generation X women takes their responsibilities seriously and appreciate brands that have a strong sense of integrity.

Provide support

Generation X women desire recognized structures to compensate for the reduction in social and cultural structures they experienced while growing up. This means brands that can offer them an opportunity to "join our club" will be extremely appealing to these women. Marketing approaches should not be too formal or dictatorial; they need to be taken up by women on their own terms. Coca-Cola, the world's leading brand by brand valuation, with a brand value of $67 billion (*BusinessWeek*/Interbrand 2006) strongly invites consumers to join their world, which projects a nostalgic vision of Christmas Past that provides Generation X women with a sense of belonging and warm reassurance. Its advertising for Diet Coke uses a different kind of tribal belonging. It shows a group of office women who watch overtly masculine men at work. This reinforces the bonding of the women and the strength of these women in reversing the typical role of the male voyeur to give women the power over the men. The workman has become their plaything.

First Direct is another brand that has been highly successful at providing Generation X women with a responsible yet individual brand construct based on openness and integrity. First Direct has redefined the retail banking metaphor as more down-to-earth and adult-to-adult rather than the old-style masculine parent-to-child attitude. This no-nonsense approach fits perfectly with somewhat skeptical Generation X women. The bank has adopted a more open approach to communications, using everyday language and minimal jargon. They have the highest customer advocacy rates of any industry, especially difficult in the retail banking sector. They have achieved this by building straightforward relationships with their customers. Their personable call center employees are calm and friendly, and, showing that they really care, inject into interactions with customers their own personality rather than the perfunctory monosyllabic responses received from many other call centers. First Direct regularly uses its own customers in advertising campaigns, often with customers simply describing their experience and relationship with their bank – which demonstrates powerful advocacy and third-party endorsement. Their advertising process is also insightful: they simply write to customers and ask if they would like to take part in the adverts and then select those with simple stories to tell about the bank. This down-to-earth approach plays very effectively to Generation X women, who dislike being overly sold to and differentiates them from other banks that use high-profile celebrities or expensive computer visualizations to convey their message.

Individualistic

In a reaction against convention and tradition, Generation X women also want to reinforce their individual style, unlike their Baby Boomer parents who are happier to fit in with convention. This means that brands need to be very clear about which subsegment of women they are trying to attract. These women will be turned off by broad-based marketing messages that seek maximum coverage. They are looking for brands that encourage and unlock their creativity and individuality. They prefer anti-brands that are not mainstream, like Muji the no-brand retailer with its pared-back aesthetic. They like clothing brands like Jaeger and online brands like Boden, which offers non-high-street fashions, or Maybelline cosmetics, with its retro UK punk chic advertising fronted by the ubiquitous Kate Moss. Bennetton and Diesel jeans have both carved out a marketing niche that deliberately rejects the mainstream market in fashion and encourages Generation X women to see themselves as an individual. Megan Van Someren, head of planning at the advertising agency Wunderman, believes that Generation X women have "high expectations around receiving easy and personalized service at all times and companies need to deliver this by talking to them based on their behaviors and personal interests" (Van Someren 2006). Gap recently created an advertising campaign called "Individuals of Style," featuring a variety of celebrities across varied ages and ethnicities wearing mixes of Gap clothing and their own personal style in TV spots. The campaign recognized each customer as an individual by highlighting accessories from their own collections worn by the celebrities. The campaign was extremely well received by consumers, particularly women.

Amazon has changed the way women buy online. Right from the start, their approach was focused on individual choice and giving control to the user, concepts that are very appealing to Generation X women. They built strong customer relationships through sharing knowledge and recommendations, much as happens in a women's word-of-mouth community. Amazon created "one-click" shopping that encouraged customers to enter into a dialogue with their business. They rewarded their customers by negotiating an agreement: let us hold your details and we will allow regular shoppers to buy and get delivery without the time-consuming re-entering of all their address and bank details. The implicit contract is collaborative and supportive rather than divisive. The customer becomes a partner in the mutual transaction. Amazon also developed a tool to help customers select books by providing recommendations based on previous purchases, and allow you to see what everyone else is buying to stimulate purchase activity. These combined service components create a powerful retail offer. The essence of this is that Amazon, despite its size and remoteness, behaves just like the local retailer that consumers trust. It remembers your personal details and makes suggestions based on what you have previously bought. This is the kind of experience that Generation X women love because it is individual and real. Amazon has combined global scale with local intimacy, a rare feat. Amazon's brand ethos is to offer mass personalization through a brand experience that offers tailored advice and a seamless, no-nonsense service.

Self-reliance for single women

Advertisers who focus endlessly on couples are missing a trick. Today an extraordinary 38 percent of all 16-to-64-year-olds are single, a figure set to grow by one fifth to 45 percent of the UK adult population by 2010 (Carat 2005). Generation Y women are still the main group of singleton women but Generation X women are not far behind and they have very specific needs. There are rich opportunities for brands that are willing to create tailored propositions for these women. This increasing trend means these women are at the forefront of redefining the female/male boundaries, attitudes and behaviors. The old image of Bridget Jones, the spinster in waiting, is no longer relevant for modern women. These Generation X women have taken the lead and are no longer willing to wait for Prince Charming to arrive. "They are looking for sex, not partners," says Michael Florence (Carat 2005). They have taken control of their relationships and are no longer seen as passive but as active in choosing their lifestyle and the brands they associate with.

Generation X women have fought hard to stand up for their equal rights and expect brands to treat them as individuals.

Single Generation X women are particularly keen to demonstrate their self-reliance, rather than being a burden on society. A recent study by media agency Carat (2005) (singledom.co.uk) described the results of their research into singletons and found that women represented two thirds of Generation X singletons. This is approximately 2 million people in the UK alone. These women are more likely to be in higher social classes AB and C1 and half of them have never been married. The brands these single women admired more than their counterparts were BT, Boots and Waterstones. All three of these brands represent solid, trustworthy and mass-premium organizations. This group of women were described as "Pippa Pans" by Carat. They have accepted their single status but are keen that they are not marginalized by society because they are not part of a traditional couple. Renee Zellweger and Ulrika Jonsson epitomize these women, who often have a tough exterior. According to the survey they are more likely to be vegetarians and not surprisingly believe that the world would be a better place run by women.

Ironic

Clearly, for women who came of age during the post-modern design era of the Eighties, using a sense of irony is a valuable marketing tactic. They may have an everyday outlook on life but enjoy an ironic "twinkle in the eye" sense of humor, rather than the slapstick guffawing that many men appear to prefer. This communicates a witty mockery that makes the minutiae of everyday life sparkle. Brands can connect powerfully with Generation X women through subtle irony in their marketing communications. They appreciate the self-conscious attitude that says overtly, "We know we are trying to sell to you but we think you'll like this product": "I hate being sold to. That's why I bought a Saturn automobile" (Wellner 2003). This openness is disarming and differentiating and allows Generation X women to feel included within the relationship rather than excluded as an adversary. Similarly, Sprite managed to increase sales from being ranked 7 to being ranked 4 in the US drinks market by being honest in their marketing; their slogan – quoted earlier – was "Image is nothing. Thirst is everything. Obey your thirst." Again, an ironic honesty in advertising that is almost too honest to be true works wonders with Generation X women. "Sheila's Wheels" is another great brand that is able to sell mundane car insurance with a sense of ironic humor. Their all-female management team were able to fine-tune the insurance proposition with female-friendly service features. These included setting their own service guidelines to guard against patronizing behavior from car repair personnel, and providing an increased insurance limit for bags, as women tend to carry a larger total of valuables in their handbags than men do in their billfolds. What is great about Sheila's Wheels is that the brand is based in the UK, but doesn't sell insurance in Australia. It has reversed the Australian male's stereotype of the "dumb Sheila" to provide a tailored and superior insurance service specifically for women. This is exactly the kind of ironic and clever manipulation of the media that Generation X women enjoy because they are included within the relationship.

Citibank fundamentally shifted their marketing strategy based on a human truth that may seem obvious. Citibank identified that wealth *per se* was not a preferred end goal – but rather how that wealth could enrich customer's lives. This insight helped them develop three key components in their banking proposition that made the brand more appealing to Generation X women. All three of these – empowerment, protection and reward – create a warmer relationship with their consumers. First, the focus on empowerment speaks clearly to women's sense of independence. Second, protection overcomes their concerns, especially about online banking. Finally, reward sends a message that makes it clear to these women that the bank appreciates them as consumers. Citibank issues "Thank You" points rather than the typical anonymous reward points of other banks. With the help of advertising agency Fallon they translated this into range of lighthearted and quirky (for a bank) campaigns that helped to make it more appealing to Generation X women: "Live richly, there's more to life than money" and, "Hoard friends. Save money." This use of irony is appealing to Generation X women because they are more cynical than women in general and therefore do not like overly slick advertising. Its clever use of an explicit selling message – "Live richly" – works satirically and is more honest than pretending that no-one wants to have a rich life.

Sandwiched between old and new

Work–life balance is a core need for Generation X women. They have witnessed their parents' strong, overriding work ethic and the younger generation's free-living fluid lifestyles. They are caught between two very different approaches to life, wondering whether to live to work like their parents, or work to live like Generation Y women. This polarization means that they sometimes hold traditional values, while also hankering for a new way of living their lives. This is unlike Generation Y women who are utterly different from their grandparents, the Baby Boomer generation. Generation Y women do not feel the guilt that Generation X women do about their shifting roles. This guilt can play heavily on Generation X women, and brands that can assuage this guilt by giving them permission to break free from convention will be highly appealing. The growth in SUV car brands like Jeep, BMW X5 and the Land Rover Freelander is a defining characteristic of Generation X. It has become a symbol of their desire to not follow the tradition of expensive sedans and demonstrates their longing for freedom in their lives. SUVs are a potent symbol of the independence desired by Generation X. Very few customers will actually drive offroad, but it is the sense that they could do if they wanted to that is important. SUVs have just become Ford's bestselling car category in the US, traditionally a position held by the humble American pickup truck.

> Generation X women feel torn between clinging on to the best of tradition while being open to the new opportunities like their younger sisters the Generation Y women.

Honda, once the epitome of blandness in the middle market, has become a brand that embodies the desire of Generation X women. Its case is a master class in marketing to women. The brand was perceived as reliable and functional and needed to find a new way to connect emotionally with its target audience and differentiate itself from competitors like Toyota and Nissan. There were two key marketing strategies that helped Honda reinvent itself. First, about three years ago, they made a strategic marketing decision to abandon using images of their cars in their above-the-line advertising. This looked like commercial suicide in an industry that is narcissistic about its product. However, their insight into Generation X women had identified that they already believed the functional claims about Hondas and were seeking an emotional reason to purchase a Honda rather than a functional one. This was borne out when consumers rated Honda more positively without the image of the car than with it. The consequence of this shift was that the advertising had to create a warm and emotional persona for

Honda that reflected its role in people's lives without reverting to tired clichés of shiny cars. Wieden + Kennedy (W + K) have created a long line of award-winning campaigns to help move the Honda brand out of traditional car territory and convey the experience of the car and the brand without showing the car once. These include "Cog," a campaign using rolling car components like a giant game with each component bending, tipping or rolling onto the next one like a collapsing domino chain. This was followed by "Grr," an animated cartoon world with a very catchy Seventies retro song about how people hate to change. W + K's latest campaign uses an uplifting choral harmony to recreate the sounds that we associate with driving. These included the sound of rain on the roof, the squeal of tires as we drive around a traffic circle or the sound of a CD slipping into its player. Honda have not only repositioned themselves as a more female-friendly brand but as Michael Russoff of W + K described it, "We have conveyed what it feels like to own and drive a Honda," without ever showing the car. It is a landmark in car advertising and marketing that many brands would do well to follow. Similarly, in the US, Honda's "It Must Be Love" campaign linked the type of owners with their cars to emphasize the emotional connection and social identification women have with the car they drive. Both these approaches helped Honda achieve a worldwide sales rise of 5 percent to an all-time high of 3.55 million vehicles in 2006. Honda Motor Co. President and CEO Takeo Fukui speaking at the 2006 year-end news conference said, "We continue on our journey for growth; especially in the US which marks a new Honda sales record for the tenth consecutive year" (honda.com). Honda's approach shifts the emphasis from technical boy's statistics to the personal and human connection with the brand. It gives Generation X women the permission to enjoy their freedom without the guilt or tension between their desire for a new car and their cautiousness.

Volvo is another brand that typifies a shift in brand positioning designed to appeal to Generation X women. Their traditional positioning of "Safety" resonated strongly with the Boomer women but not with Generation X. Volvo's brand journey from safety to "Safe and Sexy" is a microcosm of a marketing approach that is supremely successful with Generation X women. It is underpinned by traditional values like safety but offers the freedom of being sexy as well. Volvo has created a softer, more feminine design language and its advertising is no longer based around their famous crash test dummies. They have also extended the product range to include more open-top models that paint a picture of enjoyment and freedom.

SUMMARY

Generation X women shoulder their responsibilities in life. As a group of women they feel they missed out on the boom years of their parents but were left with an economic and social hangover to deal with. The traditional institutions have broken down and society has shifted from one of structure to one of the individual. They have suffered economically and are facing a huge pension bill for themselves and for their parents' and grandparents' generations who have all lived well beyond previous norms. Healthwise, they have faced the rise of global viruses like Aids and the rapid spread of cancer as a disease that now affects one in four adults in the Western world. It is easy to see why they are more conservative than their Baby Boomer parents. Generation X women are both more individual and looking for more structure through friend and community networks. They work harder than their parents, and do not have the work–life balance of their children. They are seeking permission to break free.

Marketing to Generation X women requires being down-to-earth and real, avoiding the superficial and overly flamboyant. They are very straight-forward and skeptical about being sold to. Keeping communications authentic and grounded will appeal to these women.

Brands that encourage and support their decision in life to be an individual will be popular with Generation X women. These might be brands that encourage creativity or the relaxation of the rules that govern their structured lives. These need to demonstrate how their products benefit their total style rather than small parts of their life.

Generation X women demand real quality; they are not looking for the cheap, quick-fix solution. This means that products and services need to be well supported and back up any claims that they make. These women will check the ingredients on a food pack, or research the employee conditions of a clothing company before buying their product.

Ignoring these insights and marketing strategies will result in a very decisive and negative reaction from Generation X women who, unlike their sisters at other life stages, will not give you a second chance.

6

Baby Boomer women

PROFILE

Baby Boomer women exhibit specific secondary characteristics in addition to the primary characteristics that all women possess. Baby Boomer women are young at heart. They grew up in a time of growth and optimism. They have few debts or regrets and want to continue living their lives to the full. They are fit and active and want to avoid the decline that they witnessed in their parents.

Women born between 1946 and 1965: UK and USA (see Figure 6.1)

- 32 percent of UK female population (total UK female population 30 million. www.cia.gov)
- 9.6 million women (Carat 2005)
- 26.7 percent of the US female population (total US female population 151 million. www.cia.gov)
- 40.4 million women (Johnson and Learned 2004)

Key events that shaped their world

- landing on the moon
- Vietnam War
- the Pill
- flower power
- the Nuclear Age

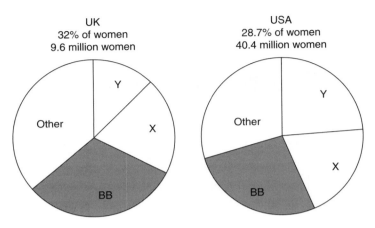

Figure 6.1   Baby Boomer women in the UK and the USA.

INSIGHTS

Women who are 50 years and older are classified as the Baby Boomer generation. They are probably married, some the second time around. Their kids are at college or have families of their own. Their health is significantly better than that of previous generations. This means that at 60 years old they look and feel like 50 or younger. They recognize their fortunate position and actively take care of their health. These women are investing in and renewing their bodies and their spiritual wellbeing. Having coped with the pressure of bringing up their children they are now free to indulge in maintaining and enjoying their own lives.

They are the richest of all of the three segments in this book and have high disposable incomes. Their mortgage is paid off; their kids have left home and they have salaries and pensions that are both significant. "The over 50's in the UK have a collective pot of £175 billion pounds of disposable income, which is greater than any other group" (Saga 2006). The figure is even higher in the US:

> The boomers' collective billfold will only get fatter as they continue working. As a group, people age 50 to 60 are flush, with more than $1 trillion of spending power a year, about double the spending power of today's 60-to-70-year-olds. (Lee 2005)

The food they eat is healthier than ever. They eat fish that is rich in Omega-3 oils that helps aching joints. They take notice of the latest food trends and take vitamin and mineral supplements. These women take an active interest in new research evidence about which foods help combat cancer, Alzheimer's and other degenerative diseases. They manage their health with low-carbohydrate food to boost their energy. They feel great and don't want to quickly wane as their parents' generation did. Boomer women in particular are a strong focus of businesses because they live longer than Boomer men. On average that is three years longer than men in China, five years more in the UK and six years longer in the US (cia.gov). In fact women in the US who reach 60 can expect on average to live an additional 23.5 years (*Wall Street Journal*, 26 September 2005). The net result is a generation of women who are active, experienced, unfettered and wealthy. They are indeed in the prime of their life. This naturally makes them an exciting yet challenging target for marketers. The biggest single difference between those companies that are successful with these women and those that are not is the ability to make them not look and feel old. In their heads they are still youthful women; and outwardly this is increasingly so too.

> [S]pecialty retailers from Gap to Gymboree are tapping into the spending power and fashion savvy of boomer women – the 77 million female Americans born between 1946 and 1964 who have long set the pace for marketers and advertisers. (Betts 2005)

Baby Boomer women grew up as teenagers during the growth years of the Sixties and Seventies. The divorce rate was low and the number of children per couple was high – indicators of a collective view on a prosperous and healthy future. This was a time of optimism and a sense of endless growth in the future. Women's role in the workplace was becoming more equable. They were moving beyond menial or temporary work and becoming important managers in their own right. This has however been a slow and difficult journey and the gender divide at work is still significant. The relationships that Boomer women joined were more unequal than today, with many women expected to bring up the children. The mother, while more emancipated than previous generations, was still the center of the family and ruled the household. In the UK there are currently over 19 million people over 50 (some 41 percent of all adults) and they represent the fastest growing sector of the UK population and control 80 percent of the disposable

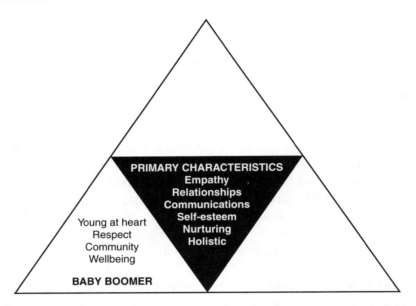

Figure 6.2 Primary and secondary characteristics of Baby Boomer women.

wealth. Women make up more than half of these and are one of the biggest audience groups in the UK (saga.co.uk). In 2000, 37 percent of the adult US population was 50 or older; by 2010 that figure will have risen to 43 percent (*New York Times* 25 July 2001). In 2001, the 50+ group held $29.1 trillion in net worth, or 69 percent of the US total, up from 56 percent in 1983 (AARP 2004), This makes them a primary business target for every sector.

Newspapers and the radio are key communications channels for Baby Boomers. They grew up with these media and these have informed their values, beliefs and ideals (Projectbritain 2005). However, radio in the UK is still dominated by the BBC radio channels, which are ineffective for advertisers. Motion pictures and the internet are much less effective at reaching this audience.

MARKETING STRATEGIES AND TACTICS

Young at heart

Baby Boomer women (see Figure 6.2), like many at other life stages, are younger at heart than their years suggest. "Thirty-five plus is code for brands that probably have an average age of around 50" (Chung quoted in Betts 2004). The outward signs of aging may be difficult to slow down but their sense of vitality is not diminished. These women recognize that this is probably their best chance to indulge themselves since they went to college. They have played the dutiful wife role, brought up a brace or more of kids and supported their husband through his career. Now it's time for them to have fun and enjoy themselves while they are still fully active. They have watched their parents decline rapidly during their final years and are determined not to let that happen to them. Luckily a better diet, healthier lifestyle and better healthcare are on their side. "My only problem is owning up to the fact that I'm really not a kid anymore, although in my mind, I'm still young," says Judith Siller, a 55-year-old real-estate broker from West Palm Beach (Brown 2006). Marketers must avoid patronizing these women. Brands should always aim several years younger than their current audience and this is especially vital when targeting Boomer women. They are not ready to be treated like old-age pensioners just yet. Their parents' generation had few opportunities as they reached middle to old age; but the opposite is true for Boomer women. They are in fact in the prime of their life and marketers need to identify opportunities for their brands to reflect this youthfulness.

Creating dedicated "older" products, services and brands will only patronize these women. Marketers need to adapt their brand messaging to speak more inclusively and engage Boomer women. They should ensure that DVD players are easy to use (something that will help all of us) or that the print on an insurance brochure is high-contrast and a decent font size for example. It is getting these little things right that will avoid alienation and will be successful with Boomer women. Shredded Wheat is a great brand that focuses on its benefits for a "healthy heart" as a core marketing message. This cleverly speaks to the older generations without labeling it as a medicine for an ailing generation. It is also a brilliant example of a message that is dynamic and optimistic; it's about taking control of your health and improving your chances of a long and healthy life.

Underscoring this young-at-heart mindset, Pepsi-Cola's 2002 Super Bowl commercial illustrated how advertisers can appeal to Baby Boomers indirectly, while targeting the 18-to-34-year-old demographic. They used Britney Spears to appeal to younger audiences but the commercial illustrates her traveling back in time, highlighting the good things about previous decades. Featuring the then popular teen icon, the commercial depicts scenes from the Fifties, Sixties and Seventies, with young people drinking

Pepsi in a relaxed, carefree atmosphere. While the commercial appears to target predominantly younger audiences, the tagline, "For those who think young" appeals to the attitudes and mindset of Baby Boomers.

Sixty is the new fifty for these women. They have a youthful mindset and are full of life and energy.

Indulge me

These women have a mature confidence borne of experience and now have the financial means and opportunity to do what they want. Unlike their parents, who were often worried that their pension would not be enough to live on, the Baby Boomers have excess cash compared with Generation X and Y and they want to indulge themselves before it's too late. Brands need to show these women opportunities for indulgence that go beyond the previous bounds of their world. They need to inspire women and help them expand their world of possibilities. These brands need to give Boomer women the confidence to have a go and try new things. With encourage-ment they will indulging themselves by travelling in business class on their long-haul vacations for example. This kind of profligacy would have been unheard of by previous 50-year-old travelers. One of the many differences of Boomer women is that they are the first generation to "SKI": Spend the Kids' Inheritance. They have seen their children grow up, get married and build homes and families. They no longer feel obliged to leave them all their wealth to aid their lifestyle, but rather enjoy what they have worked hard for themselves and see themselves as deserving to enjoy while they still can. These Boomer women also want additional services rather than just products. They have a wardrobe full of clothes but are unsure how to put them together. The US retailer Nordstrom has recognized this issue and responded with style consultants to help their customers choose and coor-dinate their look. This is part of a long-term marketing plan that focuses on building longer-term relationships: once a women trusts you with her look once, she will be back each season for a refresh.

When Vespa relaunched in the US its marketing focus was twentysome-things looking for a cheap, fun ride. However, it is the Baby Boomers who have surprised the executives at Vespa by buying a quarter of sales each year, and that's without even targeting them. Vespa believe that by focusing on them they can raise that figure to one third of sales (Lee 2005). The reason these women are unexpectedly attracted to a Vespa lies in its cute, candy appeal that is reminiscent of their childhood. The postwar freedom that came with cheap transport was a crucial step towards independence for a young Boomer woman. Now that their children have left home these women are recapturing that sense of freedom. Without the financial constraints of raising a family, these women are indulging themselves in the higher-end models and Vespa has increased the quality and number of additional features and accessories so not only are they a high proportion of sales, they are also the highest-value customers, unlike the cash-strapped Generation Y women. To increase its visibility with this audience Vespa advertised at the Golf Course Home Network, a website aimed at Boomers seeking homes nearby golf courses.

Making them feel part of the mainstream is key to succeeding with these women. In 2001, Neutrogena introduced a new specialty skincare product, Healthy Skin Anti-Wrinkle and Anti-Blemish Cream, designed to combat both acne and the visible effects of aging. The company targeted the product at women aged 24 to 44, with more than 60 percent of promotional spending targeted to late Baby Boomers. By developing a product and marketing that addressed a set of conditions that spanned age groups, Neutrogena successfully appealed to both younger consumers and Baby Boomer women.

Respect

The last thing these women want is to be patronized or written off as over the hill. They need to feel like they are still young at heart if not in years. For too long, advertisers have depicted old age as a form of illness or a disabling condition. This makes women feel like social outcasts and ignores their true attitudes. Levi's ran a campaign recently that used older models to underline the key message that people never stop being a rebel, whatever their age. This message resonates strongly with Baby Boomer women as it is inclusive and aspirational at the same time. Role models and celebrity endorsements should therefore be young at heart. Revlon use Susan Sarandon in their advertisements and Marks & Spencer use Twiggy the original Sixties fashion icon in their clothing advertisements. These women demonstrate energy and active roles in society as well as in reflecting their customers' relationships and personal lives. The communications language should be straightforward and direct; these women have witnessed many changes in their world and expect brands to cut through to the issues and benefits directly. They are still no-nonsense people. A great example of showing these women respect is the Estée Lauder campaign. In order to cater to the large percentage of Baby Boomers concerned with the visible effects of aging, Estée Lauder utilizes models of similar age to establish trust and convey product benefits. The use of an older model helps convey the message that Baby Boomers can age gracefully, defying the effects of aging and maintaining a youthful lifestyle and appearance.

> Baby Boomer women don't want to be patronized or labeled as not interesting and consigned to old-people-only products and services.

These women feel that it is their time to be themselves and live a little. This means new opportunities for marketers, where tradition would have dictated that brands and their communications were confined to focus on an old people's world full of stair lifts and incontinence pants. Those that implicitly portray a younger and more glamorous woman within advertising imagery will reap the benefits, while these women acknowledge they may still need these products. There is no reason they should be portrayed as disabled. Mitsubishi Motors mistakenly changed their target profile to 20-to-35-year-olds in the belief that youth was always desirable. While their average age of buyer dropped from 40 to 35 in two years the financial result was less positive. These people simply spent less money on their cars and their core customers, the 40-to-55-year-olds, were turned off by the Indie music and edgy advertising imagery. They retuned their advertising with fortyish actors and more realistic imagery and recaptured their most profitable segment, the Boomers (Lee 2005).

Community

Like all other women Baby Boomers strongly value their connections with others. The Red Hat Society in the US is a network of 50+ women who are not slowing down into old age. They are "greeting middle age with verve, humor, and élan" (redhatsociety.org). The society's purpose is to share stories, promote an active older life stage and help others achieve further success and satisfaction. The University of the Third Age (U3A)(u3a.org.uk), founded in France in 1972 and active in many countries, provides a more structured but equally enabling environment for older women and men. Their focus is on educational, creative and leisure activities for its members. Eons (eons.com) is another new brand. Set up by Jeff Taylor, the founder of job search site monster.com, it targets Baby Boomers with its strapline "50 Plus Everything" and is designed to "inspire a generation of boomers and seniors to live the biggest life possible." It encourages people to connect across a range of topics like people, fun, love, money, body, life dreams, obituaries (the only site that recognizes this subject in an objective manner), life map, and travel. It is a refreshingly young-looking website that demonstrates the "young at heart" principle that is the guiding principle for Baby Boomer women. What is different about this site that makes it so female-friendly is that the majority of the content focuses on people and relationships. Its home page introduces new people with cameo descriptions and has lots of features and links about the community. It is based on membership and therefore encourages users to treat the site as a daily ritual or newspaper to foster greater community feelings. The most innovative feature is the life map tool that allows users to create a visual journey through their life. In an overview mode it shows each decade with small icons to illustrate each life event such as a vacation, their health, a partner, family or child event. By clicking on these icons an image and story is revealed for each life event. The building of a personal memory map is extremely fulfilling as it helps order and provide a clear perspective on a range of all aspects of a woman's life.

> Baby Boomer women grew up when community was a reality and they are maintaining this sense with online communities.

This generation of women are truly becoming the golden surfers. Online usage is not confined to Generation X and Generation Y women; Baby Boomer women increased their online spending by 129 percent in the last two years. According to the Communications Consortium, women over 55 are the fastest-growing segment of internet users in the world (ccmc.org 2007). This of course matches the general growth in retail for Baby Boomer women; but what is surprising is that favorite sites are grocery, DIY and electrical brands (Armstrong 2006).

There are three reasons why Baby Boomer women are ideal customers for online retailers. First, they have some of the highest disposable income levels of any segment. Second, they have plenty of time to research and order their goods. Finally, they are significantly different from women in other life stages because they are available for home deliveries far more, making e-tailing an incredibly suitable medium for these women. For marketers it is crucial that their websites are easily navigable for these women who may be beginning to experience weaker eyesight, hearing or dexterity. It is simply good practice to design for an inclusive society and marketers are ignoring the biggest going if they do not build sites with these women's characteristics in mind.

Businesses that help strengthen and facilitate a women's familial and social network will be well rewarded. Travel companies like British Airways or Travelocity use their online check-in facility to make it easier for women to research, book and visit their grandchildren. Others offer tailored packages for groups of women, like the China Tour which combines seeing the classic tourist sites like the Forbidden City in Beijing and the Great Wall with visits to local silk-makers and clothes designers as well as flower markets and cookery schools. These women are looking for genuine and authentic travel experiences and will heavily research their options before deciding which brand to book with. Providing detailed itineraries and examples of hotel rooms, facilities and feedback from other travelers will all encourage women to trust and book with your brand. Women have always been extremely active socially and prefer the human touch when it comes to service. Traditional banks that are able to recognize and reward their longstanding female customers will benefit far more than the internet banks (even though women are high internet users). It is the personal contact and sense of relationship that keeps these women returning to the banking hall or retail outlet. They simply prefer and enjoy the experience of social interaction.

Recognizing a need to remain engaged with their core customer, the affluent, aging Baby Boomer, retailers such as Saks Fifth Avenue and Lord & Taylor revamped their store layouts to tailor to this demographic. For example, L&T implemented comfortable, round-armed chairs throughout the stores so older spouses and friends can relax while shoppers try on clothes. Saks Fifth Avenue stores also feature refreshments for customers, emphasizing outstanding customer service and larger, more comfortable fitting rooms containing chairs.

Health and wellbeing

Baby Boomer women are healthier than ever for their age and have the time and interest to invest in their bodies to ensure that they are in tip-top condition. 32 percent of over-50s exercise more than four hours a week (12 percent more than under-50s) (saga.co.uk). Life expectancy in Westernized countries is high and even at 65 years old, women can expect to live well into their late eighties and nineties. Although many women will need care as they get older as well, 50 percent of all women over age 65 will require some form of long-term care (MetLife 2005). Even in Asia, life expectancy for older women is high and rising. Between 1990 and 2001, life expectancy for Asian women increased on average by 3.4 three years to 70.8 years. This is a massive 26.3 percent faster than men, whose life expectancy rose by 2.7 years from 63.2 to 65.9 years old – almost five years less than women (Asian Development Bank 2004).

American Baby Boomer women were the first generation to benefit from a piece of sports legislation called Title IX that became law as part of the 1972 US education legislation. It required part or fully federally funded schools to provide girls and women with an equal opportunity to compete in sports. In the following three decades this has had a profound impact in rebalancing women's engagement in sports in both absolute numbers of women as well as the number of women's teams within each National Collegiate Athletic Association (NCAA) member's campus (see Table 6.1).

There is a quirky Californian sportswear brand inspired by this legislation called TitleNine. It has an evangelical mission to encourage girls and women to participate in individual and competitive team sports. The brand is based on a strong moral cause that is synchronous with that of Baby Boomer women. TitleNine sportswear is designed to provide increased competitiveness without the loss of feminine style. The site is easy to navigate and offers a range of high-impact and high-flexibility sports from rock-climbing to yoga. American Baby Boomer women's interest and current state of health has clearly benefited from that change in law.

Table 6.1 Effect of Title IX on US women's engagement in sports (Source: titlenine.com).

US women sports participants	1972	2007	% Increase
Number of high school girls competing in sports	1:27	1:2	1000
Number of female college athletes	32,000	151,000	500
Women's teams per campus	2	8.3	400

As well as food products that can enhance their wellbeing, vacations, leisure and other relaxation activities can all benefit from targeting older women with an inclusive message. The most effective way is to stimulate and encourage these women to a better lifestyle. "62% of Boomer women say they work at trying to maintain a natural appearance" (Brown 2006). The Elemis Spa brand typifies a focus on the active health and emotional benefits of wellbeing. The Kiehl's brand is another great example of focusing on the benefits of quality skin and hair care and their packaging is exactly the kind of instructive and non-nonsense design that will appeal to Boomer women. Both these brands acknowledge that these women are not interested in superficial advertising claims and that they have a witnessed a lifetime of "snake oil" solutions to the aging process. They are also not interested in campaigns showing twentysomething waif-thin models trying to suggest that if they too use the moisturizing product they can look like them. This is far too patronizing for Boomer women who are comfortable with who they are, what shape their bodies are in and even their laughter lines. Packaging designer JoAnn Hines confirms their desire: "I want to look good for MY age not the age I was 20 years ago" (Hines 2006). These women want genuine products that are realistic about their bodies and the product's effectiveness.

Pillsbury recognized the increasing number of Baby Boomers were becoming "empty nesters" and the impact that this lifestyle change had on their cooking habits. The company developed a campaign entitled "Cooking for Two," designed specifically for middle-aged couples. They launched the campaign at the American Association of Retired Persons (AARP) National Campaign and Expo in October 2004. The campaign represented Pillsbury's first-ever marketing effort aimed at helping empty-nesters adjust to buying, cooking and organizing meals for a smaller household and featured the company's line of oven-baked frozen dinner rolls and biscuits. "Cooking for Two" consisted of a television advertising campaign, meal preparation and shopping information on the company's website, a monthly electronic newsletter and a PR campaign that featured Baby Boomer and Olympic figure-skating champion Peggy Fleming, a recent empty-nester.

The Clarks shoe brand has always focused on the pediatric benefits of its range and this has been an appealing message especially for mothers who are anxious about their children's feet. This message resonates equally with Baby Boomer women who are conscious that a lifetime of wearing fashionable but often painful shoes has left their feet in less than perfect condition. Again, it is the honest reality that Clarks portrays as a brand that is seductive to these women. These women appreciate that they are not targeted explicitly with shoes for old people yet the message of fashionable, healthy, comfortable shoes is ideal and entirely appropriate for them.

> Because Baby Boomer women are enjoying life so much they want to make sure they stay in top shape to make the most of it.

While these women are the fittest of any previous generation, they know that they have a limited time before the aging process will start to catch up with them. This gives them added impetus and confidence to try new things. It also focuses their attention on their own and their family and friends' wellbeing. This is not a superficial attempt to airbrush the years away but a more genuine focus on enriching their life, soul and wellbeing. According to Mary Brown (2006), 96 percent of women rate a healthy family as their top priority. A report by Time Asia (2000) highlighted that arthritis affects more than 65 million Chinese, 150 million Indians, 10 million Japanese and 20 million Americans. This dexterity- and mobility-reducing degenerative condition is incredibly painful but there are many ways to diminish its impact. Savvy businesses are recognizing that conditions that impact wellbeing are now a major business opportunity. The key marketing message is long-term prevention rather than cure. The healthy drinks and yogurt brands Benecol, Yakult and Activa from Danone all focus on the digestive health of their product as a key selling-point. This message is not overtly aimed at older generations but it is implicitly focusing on a priority message for these women. Product ingredients and benefits are carefully researched and scrutinized by these Baby Boomer women so brands need to be transparent and honest about the real benefits. The American cosmetics brand Philosophy cleverly targets Boomer women through an inclusive approach with their line: "Real beauty should never be age sensitive." It focuses on a simple purity that has genuine depth to the proposition. The brand is positioned as informative, open and technically superior. This is wrapped in a brand personality that is clean with charming touches of retro such as the black-and-white images of bygone children dressed in their Sunday-best clothes. They have products called "Soul Owner" and there is a charming story on the bottle that reads, "Your only true assets are your values, your integrity, your thoughts, your words, your actions and therefore your destiny." This is exactly the kind of authentic message combined with a contemporary design style that will appeal to Baby Boomer women because it does not target them explicitly yet the underlying message appeals directly to their needs and desires. These women will not be persuaded by a slick advertising campaign. They also do not want advertisements that focus too clearly on them as a minority or specialty audience.

Kellogg's is another brand that has focused on healthier ranges for the population in general and the Boomers' desire to live a long and healthy life. They discovered that these Boomer women were becoming amateur nutritionists and with the help of the web were becoming extremely well informed. Kellogg's micro site healthybeginnings.com is dedicated to being healthy and includes nutritional information about fibers, grains and digestion. It also has a range of tools to help assess and maintain a healthy body. These include a body mass index calculator, heart health assessment and a fiber assessment. External links to the American Heart Association's "Go Red for Women – Love Your Heart" campaign help position Kellogg's as a leader and authority in the market. They have also developed a new product called "Smart Start" to add to their other healthy products like All-Bran and Special K to help stay in shape. There are three varieties of Smart Start focusing on antioxidants, soy protein and a healthy heart. Kellogg's used direct messaging that was designed to appeal to Boomer women who do not want to be conned by slick marketing campaigns. The line "More and more women are hospitalized for heart disease" has been successful and sales have risen 48 percent from a year ago while the general cereals category has reduced by 0.2 percent (Lee 2005).

SUMMARY

Baby Boomer women are very savvy about their attractiveness to businesses. They grew up in the Sixties and are the first real media-aware generation. They know they have some of the highest levels of disposable income of any segment and still have the physical and mental ability to enjoy their lives to the full. This means businesses need to target them without over-targeting or patronizing them. Along with the primary characteristics that differentiate women from men, Baby Boomer women demonstrate a youthful zest for life, require a show of respect, like to retain a sense of community and are keen on maintaining their wellbeing. All these characteristics will heavily influence marketing strategies and techniques.

Baby Boomer women are young at heart, which means that at 50 they feel like 40; at 60 and 70 they still feel active and middle-aged, rather than old-aged. For marketers this means it is essential to avoid patronizing these women or trying to fit them in a stereotypical marketing description. These women are as individual in their fifties as they were in their youth.

Because they feel and act younger than previous generations, Baby Boomer women still want to be indulged and to indulge themselves. This might be through looking good or trying out new things. Because they grew up in an era before cheap and easy long-haul travel, many of them are making up for lost time with multi-month-long vacations to Australia, the US and China.

Baby Boomers have watched their parents grow old and incapable and they are keen to maintain their health and mental wellbeing. They take an active interest in the food they eat and take regular exercise.

Finally, they want to retain their sense of community in an increasingly connected world. They will use email and mobile phones so that they can stay in touch with their grandchildren and children who may have moved to big cities or followed their careers abroad. Brands that want to build strong relationships with these women need to take these things into account in order to succeed.

7
Marketing communications

There are significant differences in the way women communicate and respond to marketing messages compared with men. This chapter identifies the key differences in marketing activities required to communicate more effectively with women. First there is a detailed account of the female-friendly adaptations to the core tools of any marketer: the language, tone of voice and visual style of communications that are more appropriate for women. The second part looks at the most effective marketing strategies and messaging themes that will be more compelling to women. It identifies the key touchstones that help persuade and influence women shoppers.

OBJECTIVES

- Identify the differences in marketing tools like language and style that are more attractive to women customers
- Define the marketing strategies that are more persuasive to women

MARKETING TOOLS

Two genders divided by one language

Women are better at conveying and understanding emotional communications while men are better when being literal. Men use words to be factually accurate, rather than emotionally engaging or descriptive. When developing brand communications it's important to recognize these two audiences have widely different expectations and understanding of the actual words being used. Women will treat the factual marketing messages as merely suggestive and indicative, while they will be persuaded by emotionally rich language that describes how they might feel. Historically, women's and men's different language abilities have been characterized as respectively weak and powerful. The power nature of the debate has emphasized strength against weakness, not difference and equality. The modern interpretation of this, outlined by Tannen (1990), is that both women and men are bilingual – they both understand and use each other's language system, but in same-sex groups will revert to their own. A woman's linguistic ability to speak in a masculine way directly affects her ability to communicate with the opposite sex and vice versa. However as we have seen, women have inherently better language skills and therefore also have stronger bilingual skills than men. This means that women can communicate better with men as well as other women, while men remain poorer at communicating with women. It is vital that marketers gender-check their communications and campaigns to ensure that they are not using a male approach to a largely female audience nor a female approach to a male one.

Women have stereotypically used more deferential language than men' reflecting their relationship-based attitude. Self-help books for women that deal with language and conversation issues often exhort women to be more direct and assert their views while underlining men's lesser language skills. Women are more articulate, make fewer language mistakes and have a wider vocabulary than men. This may inform the stereotype that women can't stop talking, but the reality is that they are simply better at communicating than men. There are several types of verbal statements women use to build stronger relationships with others (see Table 7.1).

Women, with their greater empathy and interest in others, often ask questions more than men, in order to express interest in the speaker. They have historically been encouraged to be more tentative in their tone to avoid sounding aggressive. The final key difference, one that Margaret Thatcher the British Prime Minister overcame, is that women's voices tend to be more high-pitched and therefore perceived as less authoritative. Thatcher overcame this through speech training and later television clips of her show a slower, deepened voice that has a smaller range of tonality. Thatcher was voice-coached to increase the perception that she carried strong authority. Being Britain's first woman Prime Minister meant that she had to overcome huge historical masculine prejudices, her voice being one of them. Research has consistently shown that women with deeper-pitched voices are considered more intelligent and authoritative (Pease 2001). However,

Table 7.1 Types of verbal statements women use to build relationships with others.

Taxonomy	Description	Example
Disclosure	Statement of personal emotions or attitudes	I'm bored
Question	Request information	Do you like coffee?
Edification	Statement of information from an objective source	It is Wednesday today
Acknowledgement	Recognition of another thing or person	You have arrived
Advisement	Suggestion to someone else	Come here
Interpretation	Rephrasing another person's statement	Your tone suggests you are angry
Confirmation	Starting agreement or disagreement with others	I think your tie is lovely too
Reflection	Summarizing another's statements	You said you wanted a black coffee

in contemporary society the emulation of masculine standards is no longer seen as the ideal. British Airways recently researched its on-board safety video voice-over to check what kind of voice was most appreciated by passengers. They discovered that a female voice projected the right level of authority and care while encouraging the most passengers to listen.

Women use language as a tool to share feelings and build inclusion, while men use language more to establish hierarchy and distinction.

People are more likely to read a text written by someone of the same gender as themselves. Literature has long been divided by the gender of the author. Women often read books by either female or male authors, while men rarely read anything by female authors. Like any book, the personality of the author is highly visible. In fact most books say as much about the author as they do about the subject. Female authors write in a way that reinforces the reality that women have a greater interest in social, emotional and human aspects of their lives. Moss (2003) analyzed the Orange women's book

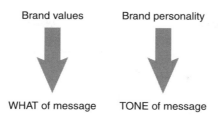

Figure 7.1 Derivations of "what" and "tone".

prize to identify if there were any differences in criteria for a good book between the female and male jurors. Overall their choices were significantly different. The women jurors placed greater emphasis on books that had a comment on life, while the men jurors preferred the idea that a book is a piece of art, more abstract.

Tone of voice (TOV)

55 percent of communication is the way you behave when speaking; 38 percent is your tone of voice (TOV) and only 7 percent is from the words you say (quoted in Heathfield 2008). Women are much better at articulating their feelings, which means that an emotional TOV is more aligned and will be more effective with other women consumers. Men typically prefer a more rational or factual TOV. A tailored tone of voice can dramatize marketing messages with an emotional personality that will demonstrate to women that the brand has understood them and is on their wavelength. Equally, the emotional range can be much greater when communicating with women than with men. Women have a greater emotional vocabulary, capacity to decipher and appreciate additional emotional variety. They do not like communications that are emotionally void or flat as they perceive these to be lackluster and unappealing.

Defining *what* your marketing should communicate is defined by the proposition and brand positioning. The *tone* used should be derived from the brand personality. It is important to balance these two components (see Figure 7.1). It is no good having a very seductive message that is conveyed with a ministerial tone or an authoritative message communicated with an ironic tone of voice. Too often businesses fail to recognize that TOV can have more impact on what is conveyed by their communications with women than the wording of the message itself. A strong tone of voice will create a truly engaging and memorable communication to your customers.

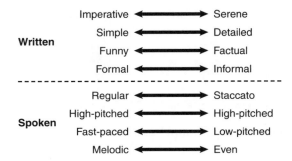

Figure 7.2 Key dimensions of tone of voice for written and spoken messages.

Women are able to understand a much wider melodic range than men. This means that a rich TOV will be more persuasive to women than it will be to men. Men with their relatively unmelodic range will struggle to comprehend complex tone-of-voice communications and will therefore be less effective.

Defining tone of voice

As we have seen, TOV is based on a number of different dimensions for both written and spoken messages. These dimensions represent the main characteristics that should be considered during tone-of-voice development. There are eight key ones, four for written tone and a further four for spoken TOV (see Figure 7.2).

In order to define an effective TOV for your brand, a precise definition of the target audience is required. For teenage segments or Generation X women this is particularly important as these are some of the most skeptical of audiences and any attempt to patronize them will be swiftly ignored and derided. It is advisable to benchmark competitors' TOV to assess preferred directions. Some brands differentiate themselves from the competition more by their use of TOV than by their use of visual imagery. Virgin Atlantic uses irreverent humor to express the brand's popularity and cheekiness. They convey that even business travel doesn't have to be stuffy or oldfashioned. This contrasts directly with British Airways' formal language and colder image. *The Economist* has both a distinctive visual style, but their TOV, insider jokes are even more memorable and engaging (see Figure 7.3).

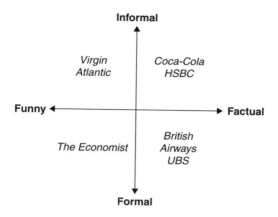

Figure 7.3 Examples of brands differentiated by tone-of-voice dimensions.

Humor is one of the most difficult dimensions to define for a brand because it has so many subtle variants. It is the easiest dimension to get wrong with women. Humor has a wide and distinctive range from the belly laugh, slapstick and obvious funny to the wry, dry and ironically witty. The benefit of using a humorous tone is that it is highly engaging, and creating a smile on a woman's face is a sure way to get them to like your brand. However, overusing humor can just as quickly turn women off, especially if it is labored or dogmatic.

London Underground recently reviewed their TOV strategy and defined three levels of communications. This was based on how directive or informational their message was required to be. These were TELL, PERSUADE and INFORM. The first level of telling people was required when a message needed to be obeyed, for health and safety or efficiency reasons:

- mind the gap
- dangerous power cables
- fire exit

The second level was used to persuade people to use the tube more effectively:

- allow elderly or pregnant people to use the seats
- use all the space in the carriage

- an electronic pass is cheaper than a paper ticket
- try a bottle of water in the summer heat

The third level was designed to inform travelers of the options available to them and encourage them to use the Tube more often:

- open till late at weekends
- Kew Gardens is lovely in the spring
- use your electronic pass on buses as well as the tube

This format gave the copywriters at London Underground a clear direction for the different TOV they needed for their varied communications tasks. One of the characteristics was to use a lacing of the Londoner's native wit as the basis for their communications to bring an English sense of humor to passengers' journeys. This lightness of tone ensures that what could have been a cold and official message is conveyed with a friendly and characterful personality that becomes a warm memory of a Tube journey.

Color

Research has shown that women are more sensitive to color differences than men. They are more influenced during decision-making than men and have a greater color vocabulary than men. An experiment undertaken by Gloria Moss and Andrew Colman highlighted that women were both more interested in color and more adventurous in its use than men. Their research analyzed the use of color in the design of business cards by both men and women. Different cards were analyzed and controlled for other external variables. 47 percent of women used colored cards or colored printing compared with only 26 percent of men. There may also be physiological reasons why women are better with color than men; for example it is notable that only 0.5 percent of women suffer from color blindness compared with 8 percent of men.

Color is one of the most powerful communications tools to use with women because they prefer colorful visual content more than men. It is also the most recognizable and memorable part of a brand identity. A woman may not remember the exact brand name or shape of the logo, but she will almost certainly remember that whether it was blue, red or green colored (in fact over 50 percent of logos are blue). When color is used, it needs to be used in a dominant fashion, like the Orange mobile phone brand or

Coca-Cola with its dominant red. Each "owns" their respective color in their market sector. Categories tend to have their own color palette; butters and margarines tend to be blue and yellow and washing powders tend to use green. A new brand identity refresh program needs to take into account the category color and decide whether to go head to head with this or go counter to it to increase standout and build its own identity with a very different color scheme. There is a strong psychology to the use color, but it is highly dependent on local cultural influences. So for example the color red often means danger in Western countries, but is considered a sign of good luck in certain parts of Asia.

> Women are more sensitive to a finer range of colors than men, who tend to find sharper, contrasting colors more appealing.

There can be issues when a brand that "owns" a particular color is challenged by another brand with a similar color entering the marketplace. This occurred recently in the UK when the Orange mobile phone brand (owning the color orange in that market) was challenged by easyMobile (a spinoff from the Easy Group's easyJet business). Since easyJet also own the color orange in their marketplace (airline travel) it is clear why they wished to continue using the color when they entered the mobile phone market. The key question for trademark lawyers is whether there will be confusion in the mind of the customer over the two offers. The courts are reluctant to give color ownership to brands such as Orange since there are so few primary and secondary colors and they are in constant common usage. In this case they rejected Orange's claims for ownership and easy Mobile has been allowed to continue its use of the orange color at the same time.

Design style

Design, like any creative activity, is a highly personal expression. The most obvious signal of this is that when asked to draw a person, women almost always draw a woman and men draw a man. Women and men are more likely to be attracted to designs created by their own gender. Moss and Colman (2001) conducted experiments to match Christmas card designers with female and male preferences. Their results demonstrated with a high statistical significance that the women chose the female-designed cards more than three times more than the men, while the men chose the male-designed cards more than twice as often as the women. There are several reasons why this might be the case. The women, with poorer three-dimensional abilities, may have preferred the two-dimensional version of the pictures and vice versa, while as noted above women's greater interest in color would probably be more attractive to them. Further research by Moss (2003) on public transport systems highlighted the gendered design preference differences of women and men. Overall the women preferred brighter colors, detailed surface designs and curved lines, while the men preferred straight, rectilinear shapes, darker colors and plain surfaces. This plainly has strong implications for marketers and advertisers. The more they consider the gender of the audience and creator as part of their communications process, the more likely they are to be effective.

Consistency is one of the most valuable tools a marketer can use in communications. It is often at odds with advertising creatives' or designers' wishes, but is an extremely effective business tool. This is because women are bombarded with thousands of images every single day, so a high level of consistency is required to build effective recognition and brand memories. With consistency comes simplicity for women. It is easier for them to recognize their favorite brand on the supermarket shelf, or favorite restaurant as they drive down the high street. The key is to identify one or two strongly visual elements and be absolutely fastidious about their execution. These are called non-negotiable design standards and the best brands like Orange and American Express protect them with fanatical zeal. Generation Y women are particularly image-driven and are sophisticated both at reading images and at using them to demonstrate their self-identity.

MARKETING STRATEGIES

When the market is overly cluttered with brands, it is often better to tack against the prevailing wind – to zig when the competition is zagging. This can be highly effective and using mind maps marketers can uncover latent connections and achieve great standout. Challenger brands often use this technique to beat far larger brands. They enter the market with what seems to be a lesser brand positioning, but use their understanding of the irregular nature of the mind map to highlight an untapped connection (see Figure 7.4).

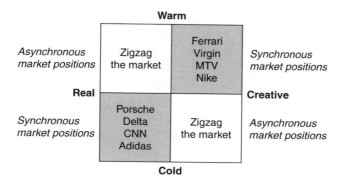

Figure 7.4 Examples of positions in possible zigzag marketing strategies.

The art of persuasion

The choice of marketing strategy needs to respond to the needs of a specific segment of women. As we have seen, there is a wide variance in marketing appreciation between Generation Y and Generation X women. The former have a huge appetite for a luxuriously mediated world, while the latter hold most marketing in disdain and are particularly cynical about being sold to through communications. However, all segments are open to persuasion if the right marketing strategies are employed and women are no different in this sense. Marketers must first acknowledge that women are very sophisticated in consuming marketing messages and can often easily decode their underlying meaning. There is a fine balance between convincing a woman and overplaying the message so that they recognize they are being targeted. Linguistic researchers suggest that because women have superior verbal skills they also have a higher comprehension of persuasive messages as well (Eagly and Warren 1976). They also assert that there is a high correlation between comprehension and persuadability, meaning that those better able to comprehend adverts are also more likely to be influenced by them. Persuasion relies on influencing people to believe the same things as others or behave the same as others and buy a specific brand.

Female purchase decision-making processes

There are marked variances between the ways women and men make purchase decisions. These offer marketers and advertisers clear insights and direction for increased effectiveness. There are two key drivers that have an impact on the design of communications that may at first appear contradictory. Women are better at seeing the big picture than men. They will piece together an overall message or impression based on all the fragments of an advertisement. Men however are attracted to single headlines or individual messages. But unlike women, they tend to be highly selective in their processing capacity. Joan Meyers-Levy (1988) in her selectivity hypothesis suggests that men make decisions based on highly selective parts of an overall advertisement. Their choice of which pieces to consider in determining the overall message is based on their own cultural understanding of what is important. In contrast, women will be more interested in reading all the detail of the advertisement and using all the facts to make a more complete decision. "Give a woman a holistic reason to rationalize her purchase and she won't think twice about buying it" (Rethink Pink 2005). This evidence echoes other research that suggests that women are more rigorous in their purchasing process. They will investigate and consider all the variables before making a purchase. The implications of this are that in order to satisfy women shoppers, it is better to have more product information available and expect that the shopper will take longer to comprehend this and then make the purchase. The conclusion from this is that women are simply more thorough processors of marketing communications and sales information than men.

> Persuading women consumers is a combination of encouraging indulgence without the profligacy that might engender guilt.

Leveraging cognitive dissonance

Women, like men, try to maintain consistency across their attitudes, values and behaviors. One of the key ways marketers persuade women is to play on the dissonance or conflict between their attitudes and behaviors, often called cognitive dissonance. For example, a woman may hold strong beliefs about her status and ambition. Marketers can play on this to emphasize dissonance by suggesting that successful people wear expensive clothes like Gucci or Prada. If that woman does not own any of these brands then she may have to reassess her values in one of three ways. First, she could admit she is not successful and change her value system (very difficult and unlikely). Second, she could acknowledge that the advertisement

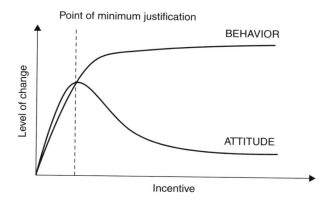

Figure 7.5 Limitations of incentivization on driving continuing behavioral change.

is deliberately trying to force her to buy a Gucci dress and overcome the dissonance by recognition of this fact. Or, lastly, she could buy the Gucci dress and complete the circle of dissonance, confirming that her success is demonstrated by the ownership of a Gucci dress. When used insightfully, emphasizing the cognitive dissonance through messaging is a highly effective marketing tool. One of the most typical ways to change attitudes and behaviors is to use sales incentives. But as Lidwell *et al.* (2003) discovered – rather counter-intuitively – there is not a direct correlation between ever increasing incentives and exponentially increasing attitudinal or behavioral change. A minimum level of justification (incentive) is all that is required, after which attitudinal change drops off while behavioral change remains fairly constant (see Figure 7.5).

Marketers can leverage dissonant situations for women so that they are driven to realign their attitudes and behaviors through consumption of new products and brands. The marketing message must therefore describe and highlight a dissonant fact or emotion that will resonate with the target women. This is demand creation and knowing that women are more persuaded by emotional messages it is straightforward to identify emotional needs and desires for those specific women. The product will then be perceived as the obvious solution to help the women resolve that dissonance, providing comfort and harmony. Hugh Rank developed a powerful theory of persuasion, known as "Rank's model of persuasion" (Rank 1976). He identified two differing strategies for persuading an audience by either

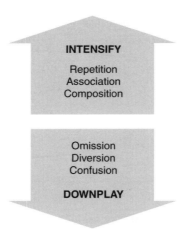

Figure 7.6 Rank's model of persuasion: either intensifying the attraction or downplaying the negatives or competition.

intensifying the attraction or else downplaying the negatives or competition (see Figure 7.6).

Intensify brand benefits

Marketers can intensify elements of a brand using three key techniques: repetition, association and composition. Intensification of the brand will increase its persuasiveness with women. First, the marketing message can be highly repetitive. Women respond more favorably to messages and ideas that they are familiar with than to new ones. Repetition will help ensure that they are familiar and favorable to your brand. Second, by linking the marketing message to memories that women already hold we can increase the intensification of the brand. These can be intensified either through attraction and positive memories or by association with negative fears. Finally, we can intensify the impact of the brand through the total composition of the brand experience. This may include combinations of the brand identity, messages, visuals, sonic branding, pace, logic and other elements.

Downplay brand weaknesses

Alternatively, marketers can downplay components of the brand to increase its persuadability. There are three ways to achieve this. First, the marketer can use omission. Identifying what should be left out of a marketing

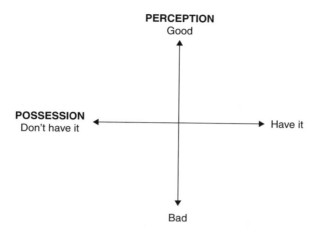

Figure 7.7 Women can reduce dissonance by actively managing their desire for things and the potential brand perceptions of that ownership.

message is often more important than deciding what should be included. Second, the marketer can use diversion. It can be highly effective to downplay the brands shortcomings by diverting women's attention. This can be achieved negatively through focusing on side issues or the trivial. It can even be by using humor and entertainment to avoid tackling the issues. Finally, it can also be effective to distract women. This can be achieved by clouding or masking the negative perceptions of the brand by creating confusion about its negative aspects, through making things sound overly complex or chaotic. There are always two parts to a marketing task: first to create the dissonance with the relevant women and second to provide satisfaction through a brand, product or service.

Express the emotional benefit to their lives

Women are constantly trying to manage a number of interrelated elements in their lives when they may encounter a marketing message (see Figure 7.7).

There are four key actions women may use to satisfy the gap between their dissonance (unfulfilled needs) and their happiness. Women's greater abilities to read marketing messages and connect complex issues mean they are more likely to negotiate across each of these four actions. They will be better able to balance the bigger picture with the detail of the marketing message. However, marketers can draw on any or all of these fundamental desires in their messaging. They can use the message within the context of their product or try to create dissonance between the brand and the customer's values and beliefs. Here are some examples of how to use each of the message types.

First, women will try to protect the good things they already have and are happy with. They may need your brand in order to protect what they already have. Clearly there is a close link here with insurance, pension and healthcare brands, but this type of message works for all products. Marketers can help persuade women to buy:

- a washing machine brand in order to keep their children happy and healthy (keep what's good)
- a brand of wine or food to protect a friendship
- a brand of car to protect their status among friends
- a brand of clothes to protect their authority in the workplace

Second, women will try to get rid of the bad things they currently have. Again there are certain sectors that fall naturally to this message: medicines, painkillers and cleaning services. This type of message can also be used to persuade a customer to buy:

- a brand of clothing to relieve or negate a negative social stigma
- a brand of cosmetics to relieve a sense of ugliness
- a brand of "loan consolidation" to relieve debt feelings

Third women will try to acquire new things that are good. Categories that naturally fall into this area of messaging are beauty and cosmetics brands but this type of message can be used to persuade customer to buy:

- a brand of car to acquire status
- a brand of beer to acquire popularity or social acceptance
- a brand of entertainment (for example, National Geographic) to acquire new knowledge

Finally, women try to avoid gaining additional bad things. Categories that align with this type of message are insurance and prevention schemes, for example:

- a brand of gym that prevents our health deteriorating
- a brand of food – for example low-sodium – that prevents us getting food disorders

- a brand of house-cleaning detergent that prevents germs growing in our kitchen

Total brand satisfaction

There are also four key strategies marketers can use to satisfy and persuade a woman their product is right for them (see Figure 7.8). These are key strategies to adopt in order to grow your business through marketing campaigns. While all four are effective in the broader market, the first, namely intensifying your own brand's positive aspects, is the one women will respond most directly to because it aligns better with their values.

Using a strategy of intensifying your own brand's positive aspects will ensure that women respond effectively because it is genuine and authentic. It is an affirmative way to build relationships that avoids being explicitly negative against competitors, while the latter, more aggressive strategy is more likely to find favor with men. The intensification approach also avoids hiding negative aspects of one's own brand that would be perceived as underhand or disrespectful towards women. Focus on your own brand with an absolute positioning and messages. This means intensify your own brand's good aspects. This usually requires an approach that deals only with your brand and product attributes, and ignores others in the market. Frequently, messages following this strategy will focus more on relevance

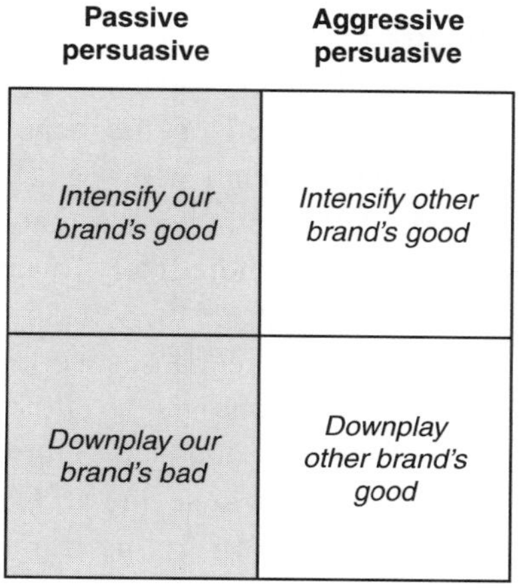

	Passive persuasive	**Aggressive persuasive**
	Intensify our brand's good	*Intensify other brand's good*
	Downplay our brand's bad	*Downplay other brand's good*

Figure 7.8 Four key marketing strategies.

than differentiation – identifying key desires and needs and explaining how your brand will fulfil this and make the customer happy. Gillette's Venus women's razor uses the theme of "divine experience" to persuade women that the product is perfect for them.

A comparative messaging strategy that is more aggressive focuses on differentiation more than relevance. Identifying successful messages used by the competition and negating or countering these will be successful. Avis's famous message, "We Try Harder," identifies their prowess as superior to the competition. The choice of marketing strategy will depend on the product category, market maturity, market share and the level of aggression or risk defined in the company's business strategy.

EFFECTIVE MARKETING TO WOMEN

The solution to creating marketing campaigns that are more effective is to align the message with the gendered characteristics of the audience. For very feminine audiences, then, strongly feminine attitudes, messages and imagery will result in more effective advertising. There is a high correlation between the feminization of the message/content, relationships, empathy, conversation and higher response rates by women. This is known as self-orientation – when the orientation of the marketing message has a high fit with one's own orientation. Joan Meyers-Levy (1988, 1991) conducted research to prove this and describe the importance of understanding different marketing messages. Using a fictitious advert for mouthwash, she tested both a "self"-directed and an "other"-directed marketing message. All other variables were controlled for and a statistical sampling technique was used. The "other"-directed message focused on the benefit of fresh-smelling breath – that is better for others around you – while the "self"-directed message focused on the benefit of healthy teeth – that is better for you. The results show that women were much more likely to appreciate the "other"-directed message than men, who preferred the self-directed message. When the gender variable was removed, the responses were similar to each other. There is a high correlation between the gender attitudes of the customer and the gender orientation of the message (see Figure 7.9).

As we have seen, women are better at seeing and understanding the big picture of an event, advertisement or marketing promotion: "Women take a broader view of life, bringing more aspects of the situation into play when making decisions" Carter (1998). They are also better, because of

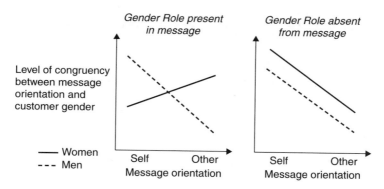

Figure 7.9   Women respond selflessly to masculine images while men respond selfishly towards feminine imagery.

their higher empathy quotient, at connecting, analyzing and comprehending the consequences of the advertisement on others in their group, be it a partner, family or friends. While men tend to focus more on a few headline elements to make their judgment, they also do this with little regard for the consequences on the wider group. These differences were characterized by Putrevu (2001) as being that women are relational processors, while men are item-specific processors. This gives strong direction to marketers who are using messaging to build relationships with their customers. Give women a big picture message first – while for men, focus on one or two key features, as men will use this as a shorthand to make decisions and consequently will not read the whole advertisement or necessarily include wider consequences. Women are more likely to exhibit detailed research and decision-making activities for product or service attributes, while men perceive the headline theme of the advertisement. There are a number of principles for designing gendered advertising messages that take into account the gendered differences in processing (see Table 7.2).

Successful advertising to women should provide a wide range of information that can be cross-analyzed to create a total picture. It will be more successful if they are verbally and linguistically rich, suiting women's stronger abilities in these areas. This might be in the form of detailed product or service information. By contrast, successful advertising to men should focus on one simple theme that summarizes the entire product or service offer. Its tone should be linguistically concise and not overly complex. As we have noted earlier, there is a direct correlation between an individual's own sense of where they sit along the gender spectrum and their

Table 7.2 Gendered differences in processing messages.

Women	Men
Comprehensive picture	Headlines only
Intuitive	Analytical
Creative	Rational
Linguistically rich	Linguistically simple
Colorful	More monochrome
Differentiation from competition	Relevance to customer
Organic shapes	Linear shapes
Associative	Objective

response to feminine or masculine advertising. Highly feminine women are more likely to respond better towards highly feminine-oriented advertisements, while more masculine women are more likely to respond positively to more masculinized imagery and copy. Clearly, gaining a greater understanding of the attitudinal target woman will help direct more-targeted marketing messages and programs.

EFFECTIVE MESSAGING THEMES WITH WOMEN

There is a discernible shift in both markets and marketing departments that suggests that the rising power of women consumers has now been fully recognized and engaged. It is not just women consumers; in fact gay men and increasingly feminized male buyers (metrosexuals for example) are more likely to respond positively to a feminine approach as well, further increasing potential audiences. Women are more likely to respond to negative emotions such as fear, tension or distress within a marketing campaign, while men are more likely to respond to positive emotions like enthusiasm, action and enjoyment. This has profound implications for how marketers create campaigns that drive dissonance and encourage product trial. For example, a campaign that reassures and builds trust is more likely to be successful with women than men. Likewise a campaign that emphasizes energy and achievement is more likely to be effective with men.

Blurring of gender roles in marketing campaigns can be evidenced for both women and men. Women's advertisements now show women as increasingly independent, assertive and professional, while men's advertisements increasingly show men as caring, emotionally mature and group-oriented. Recent advertising campaigns for diapers and household cleaning products have both used men as the role model, taking care of the children and cleaning the house.

There are six key themes for marketing to women:

- put people first, not the product – build relationships
- tell stories – they make sense of information and show how the product/service improves daily/family life
- create intimacy through warmth and engagement
- use emotional marketing themes
- show the details to build trust
- use semiotically rich language

Put people first, not the product

Women are more interested in building long-lasting relationships and this is mirrored in their view of brand relationships. They are less interested in pure transactions – they prefer a symbiotic relationship. This means that marketers must recognize that women may take longer to be convinced of a brand's value; but once they are convinced then they are more likely to be loyal. Putting people first means de-emphasizing the product features and overemphasizing the personal and human aspects of the brand. Orange the mobile phone brand exemplifies this approach. They believe that phones are about connecting people not technology, so they have always refused to show technology or products in any of their advertising. Each of their campaigns illustrates the optimism for life that is at the center of their brand. Orange launched with an advertisement showing a baby swimming underwater with the voice-over declaring that in the future it's not what you say that will be different but the way that you say it – not a technological image or product in sight. This theme has continued with advertisements like "Harmonious Dance" that show a couple so in touch with each other they ballerina around their apartment in perfect harmony. Orange has used this to illustrate the feeling their customers have when they use Orange. This is in high contrast to the masculine approach taken by O_2, who in their advertising use a technocratic style and present the product as hero. The more successful brands have strong personalities that women respond well to. These can be a character or a tone of voice or a brand personality that initiates and responds to women to create a dialogue. Building relationships, as we will see in the next few chapters, is key to building brand loyalty. Marketers need to recognize that women will respond to the opportunity to develop a relationship, so long as it provides something in return and is not cynical. In fact they will go out of their way to invest time and energy into potentially rewarding brand relationships.

> Women are more easily influenced in purchasing by people than the product fetishism that men find appealing.

Tell stories to make sense of information

Stories help women make sense of information. Women prefer to gain their knowledge through stories and emotional experiences, while men prefer facts and rational information. Women are better at encoding and decoding these and it provides a more collaborative experience that fosters relationships. It is important to recognize that they prefer not only to gain knowledge this way but also to impart it. Therefore service personnel like call center staff must be just as ready to impart emotional messages as receive them if the relationship is to grow and be fulfilling from a woman's perspective.

Connecting with a positive female character helped FedEx tell its story. After learning that women constitute the majority of frontline staff at its Asia-Pacific B2B clients, FedEx launched a television advertising campaign titled "Run Jenny Run" to better target this group. The commercial features an athletic female fast-food employee whose extraordinary dedication to customer service leads her to become a FedEx courier. Aiming to associate the company with above-and-beyond efforts to deliver on a promise, the campaign reflects FedEx's recent customer research which revealed women as appreciating and identifying with the company's "can-do" attitude when presented with obstacles (MLC 2007).

Create intimacy through warmth and engagement

Intimacy is the key to unlocking relationships with women customers. Once they have made up their minds about your brand they will maintain strong loyalty but expect intimacy. This does not mean that all brand relationships have to be highly intimate or close. It means that there is a high level of respect between them as a customer and you as the brand. The bank UBS has remarkably managed to create intimacy with its global brand. Their campaign "You & US" focuses on the intimate nature of banking. It highlights the ways its power is delivered through individuals on a one-to-one basis. This offers remarkable differentiation in the sector and demonstrates a key benefit whether you are a retail customer with the bank manger or a global corporate investment banker. As with any relationship, brands always start with a minimum level and need to gauge just how far they can develop. Different product categories have license to move both deeper and more quickly than others.

> Building trust establishes higher levels of brand intimacy and therefore persuasion.

Use emotive marketing themes

As we will see later in this book, making women feel good emotionally is the best way to build relationships with them. A brand needs to provide emotional support for women as well as creating an emotional dialogue with them. This emotional support needs to help dispel guilt around purchases and also to respond to a woman's emotional needs for support and belonging. MasterCard has recently run a campaign that reminds its card-users that their family is more important than work. It is effective because it is based on the insight that we all feel guilty about not spending enough time with our children. It shows a series of young children telling their parents that "they are fired" for not doing things with them. One of the fathers unexpectedly shows a sailing brochure to his boy and gets the response "You're hired" using the emotional connection of MasterCard with a happy childhood.

Recent research suggests that women and men manage their emotional reactions to advertisements differently. Not surprisingly, the women surveyed were more able to openly express their feelings, both positive and negative, than the men (Fisher and Dube 2005), while the men were able to express positive feelings like happiness, amusement and joy, they were less able to express negative feelings of sorrow, unhappiness or anxiety. This does not mean they did not have those emotions, only that they are more guarded at expressing them than women. This has two effects for advertising media. First, it means that women will be can be targeted with the full range of positive and negative emotions, even if their husbands or partners are watching the same television or movie advert. Feminine health issues such as their periods or bloating feelings, which often used negative images of pain or discomfort, will work successfully with women. But similar tactics will not work effectively when dealing with male health issues. This is because the men will avoid expressing their emotions, even though they may concur with the emotional message. Second, where joint decisions are required, such as purchasing health insurance, or high-ticket items, it is important to not overemphasize potential negative emotions that will turn off the male partner. Negative emotions are just as powerful and are often used to tug at the heartstrings and gain strong emotional commitment. Blue Cross, Blue Shield and Aflac, the Fortune 500 American insurance company, are effective emotional brands. Similarly, Pfizer, GSK and Eli Lilly often use positive emotions to mask the underlying negative or dysfunctional health issue. While this may be more suitable for men, women are much better at expressing and connecting directly with the underlying emotion that products like Prozac can help relieve.

Women choose brands that share a stronger than average mutual understanding of the world, a shared worldview. An advertisement for the haulage company DHL was cited by Andrea Learned, co-author of *Don't Think Pink* and well-known women's market blogger, as a great example of emotional marketing that appeals to women. She felt that the advertisement managed to build an emotional bond through a number of triggers. First, it was a situation: the blurring of personal and professional lives (the package could be for business or personal needs) is something women are better able to understand and cope with than men, who tend to compartmentalize these two areas more. The advantage of managing several tasks at once is an appealing benefit to women. The second trigger was the choice of a Burt Bacharach tune, "Love, Sweet Love," which is a "floating, light tune, and the lyrics have got to tug at most people's hearts in this day and age" (Learned 2005). Finally, the pace of the storyline is real and smooth, rather than the turbocharged version that would appeal more to men. The core insight is that women prefer to be shown the benefits of exactly how this fits into their life, rather than the overfeatures of the service. This level of insight and understanding comes through dialogue and negotiation with women; it is not something that can be prescribed in a manual. It is the essential purpose of marketing communications with women. The dialogue needs to emphasize feedback and flexibility in order to convince women that there is an intimate and live emotional connection.

Show the details to build trust

Brands must show women that they trust them and they can be trusted. First they must show themselves to be trustworthy by exposing all the details. This requires full disclosure; women are fastidious about details and will fully research a new brand prior to trial. They will want to know that the company has no illegal working practices, has an ethical stance on the environment and pursues women-friendly working policies. Waitrose the supermarket uses the detailed provenance of its produce to demonstrate its brand quality. By focusing on the details it provides rich proof points like its advertisement for local produce (which must be within 30 miles of the store). It highlights an apple juice producer who hand-delivers regular fresh supplies of bottled juice in the back of a Land Rover. It provides strong credibility and an emotional link to quality produce and the community. Limiting the information that is available about your brand will turn women customers off. They will assume you have something to hide and will find it impossible to trust the brand. In the age of the rising female customer, openness is the only way to ensure that your brand is successful.

Use semiotically rich language

Marketing messages received by women are derived not just from the words but from the cultural meaning that is associations of those words, symbols or images. Women, with their more sophisticated linguistic ability, are much more appreciative of semiotic messages. Ferdinand de Saussure, Roland Barthes and Jean Baudrillard developed the study of this kind of symbolic messaging in the study of semiotics. Their work maps out the cultural connections that women make between images and their meaning beyond the text. It relies on the fact that the words and images we use have a cultural connotative meaning beyond and prior to their textual expression. The key to both unlocking and using semiotic analysis and messaging is to recognize that it is grounded in previously ascribed meanings. The meanings are gendered and predetermined by society. They are not permanent however and can be subverted, revised or promoted differently in the future. This suggests, as Barthes does, that we must encode our world in order to experience it. Or, put another way, that we cannot experience our world without encoding it in meaningful signs. Clearly, brands play a huge role in helping women to both encode and navigate those coded signs. In this instance, brands act as codes that simplify the navigation process.

They provide bite-sized chunks of meaning that can easily be digested by women. Brands actually go beyond the realms of simple language and provide women with powerful myths that help construct their self-identities and their daily experience of the world.

- they wake up with Sony
- they breakfast with Kellogg's
- they travel to work with Renault
- they work with Microsoft
- they watch films with Disney
- and go to bed with Cadbury's hot chocolate

Women's lives are ascribed as much by the brand experiences as the objective reality of the functions of each of the items above. Truly great brands recognize that more of their role is to present and engage women in a mythical brand experience, rather than simply convey facts about product features. Every sector has brands that have become a socio-cultural myth like Disney, Starbucks, Chanel and Dove.

SUMMARY

Women have higher language and visual communications skills than men. Marketers and advertisers need to ensure that their communications are gender-tuned and checked to ensure they will be effective with their target women. Women are more holistic and forensic in their research when they are choosing a brand. This means that the level of information within marketing communications needs to be both higher and better integrated than for men, who cherry-pick headlines only.

Women's greater linguistic and communications abilities mean that marketing communications need to be more precise but can also be richer, more complex and subtler than those used towards men. Tone of voice plays a large part in the impact of above-the-line communications with women.

Research has shown that women generally prefer more colorful, complex and organic visual graphic designs than men.

Effective and persuasive messaging with women can be created by focusing on the emotional content and the relationship aspects of a brand rather than on key product features.

8

Effective brand experience design

Making women feel good is the key to them enjoying and loving your brand. As we have seen throughout this book, women value feelings and emotions more than men. This chapter defines the most effective way to make women feel good about your brand. It identifies the key tools that can be used to create enhanced brand experiences that are more appropriate for women customers. Women have a more holistic understanding of the world than men and this means that brands need to have a multidimensional approach to satisfying their needs and making them feel good.

OBJECTIVES

- Define the key moments of truth on the customer journey
- Define the ideal feeling of the brand experience
- Demonstrate the key components of the six-step feelgood framework

Women are concerned with their feelings much more than men. They want to feel good about everything they do, while men want to achieve something in everything that they do. The difference is that women are just as interested in the way things are experienced as the result of that experience. The way they feel is often the key part of the relationship they have with brands, rather than the functionality that they provide. The relationship they have with their service provider is as important as the delivery that they provide. They prioritize the way that things are done compared with what is actually done. Women are also more concerned than men that the relationship they have with a brand and its employees is a true partnership, rather than one that is purely transactional.

Making women feel good requires a different kind of effort than simply achieving their goal. Brands need to shift the emphasis to the emotional aspects of the brand experience in order to satisfy women and make them feel better. Because of the emotional and subconscious nature of our feelings, it is a complex and personal challenge to make someone feel good. However, when we feel good it is the best thing in the world. We feel more confident, satisfied, sexy and happy. For marketing managers, achieving this kind of response from our customers will unlock powerful loyalty. In order to create emotional brand experiences for women, marketers need to express feelings rather than objects or tasks. This may seem simple but is very difficult to implement. Certain brands base their positioning on a feeling rather than an achievement. Orange, the mobile phone brand, is all about a feeling; it has a warm, maternal feeling that connects powerfully with its customers. The functionality may be the same as other brands but customers recognize that they feel the brand as much as see it. Creating emotional experiences means that they need to convey a feeling, mood or sensation far more than a tangible activity.

BUILDING RELATIONSHIPS THROUGH THE BRAND EXPERIENCE

The most important aspect of marketing to women is that businesses should focus more on building stronger relationships that go beyond transactional purchases. "Time and again research confirms that women buy based on the relationship they forge with the brand or service. If you ignore that as a company, you might as well save your marketing money" (Denny 2006). These relationships ensure a higher level of loyalty and a greater lifetime customer value. Deliberately marketing to women helps businesses prioritize the choice of channels and their relative influences rather than a complete redefinition of all current customer touchpoints. Within different industries different degrees of change are required to maximize the gains. B2B and B2C businesses can gain equally but in differing ways. The heart of engagement with women is shifting from the transactional and product-based to the relationship and service-based (see Table 8.1).

Women have a much stronger propensity than men to build relationships rather than survive on a series of discrete functional connections. They are also better at building strong, long-lasting relationships. The key difference is that brands can become friends with women. There are two key elements to this context. First, women prefer long-term relationships, as these allow

Table 8.1 Seductive brand experiences require new, relationship based characteristics that go beyond the old transaction elements.

Old	New
Transactional	Relationship
Product	Service
Discrete	Continuous
Unemotional	Emotional
Non-negotiable	Negotiable
Standardized	Personalized

them to be more secure and reduce the continual effort of new transactions. Second, long-term relationships both require and deliver emotional benefits that are a higher priority for women.

This is because "[w]omen validate themselves through their relationships while men validate themselves through their achievements" (Danis 2005). As we all know, women are much better at being friends with each other and brands that can achieve this status will be rewarded with a lifetime of customer loyalty. Businesses that achieve this level of brand friendship will benefit from lower marketing and customer investment costs. Clothing champions like L.L. Bean and Victoria's Secret both ranked number one in their category in the recent brand loyalty index (Brand Keys 2007), while online brands Amazon, Expedia and Google have become the default page for many customers and also secured the number-one spot in their categories in the survey. Google in particular has become omnipresent, and many users have it as their homepage, relying on its speed and simplicity to drive quickly to where they want to go on the net.

Service touchpoints as key differentiators

In certain industries, there may be no magical insight that can drive a highly differentiating positioning. Brands often rely on the expression of their positioning to act as their differentiator, and this is especially true in commoditized markets. It is also true where there has been relatively little innovation within a market. Retail banking is a good example where the products are relatively similar and they rely on service to distinguish themselves against competitors. Research has shown that a customer's basic financial management requirements are for the bank to understand their specific needs and respond appropriately. This is true for the majority of customers and banks can therefore mainly differentiate themselves by their ability to deliver against these, rather than positioning themselves uniquely. The challenge for banks is that this is precisely the most difficult part of the proposition to control. It requires years of constant effort and resources to change a business with tens of thousands of colleagues. But when a bank gets it right, they can create enormous goodwill, like First Direct in the UK or Wachovia in the US and Citigroup worldwide. All have industry-beating customer loyalty and advocacy scores (Brand Keys 2007) and are high on the *BusinessWeek*/Interbrand best global brands league table (2007). Women have much stronger word-of-mouth networks to communicate the strengths and weaknesses of particular brands. News that a brand has a poor service reputation will quickly reach all the women in a neighborhood much faster than it will the men there. Conversely, this female attribute can be harnessed to generate greater awareness and loyalty among women shoppers. Good service is one of the primary decision criteria for women when choosing which stores to become loyal towards. A recent study showed that 37 percent of women chose one store over another because of higher service levels (Newspaper Association of America 2004).

> Brands can attain acquaintance, friendship status or get women to fall in love with them if they can make women feel extraordinary.

Different categories are able to achieve greater levels of brand friendship with women customers. Low-interest categories, such as cleaning fluids like Domestos, CIF and Mr Muscle, often represent a brand acquaintance relationship. As the category becomes more important to women, the level of relationship is likely to increase. We become friends with brands that help us through our daily activities. Breakfast cereals such as Alpen and Kellogg's cornflakes typify brand friendship relationships. They are faithful and reliable but we are not going to over-commit energies to them. Highly intimate relationships that remain short-term can be defined as lover-type brand relationships. We may fall in love with brands that help us through certain life events: exotic vacations by Kuoni, Abercrombie and Kent, or hotel brands like InterContinental or Banyan Tree, or spa weekends with Elemis Spa. These all emphasize high-impact, multisensorial experiences we become passionate about. Finally, there are the highly intimate and long-term relationships that we have with brands. These are the one that take on a spiritual dimension within our daily lives. For many these might be Apple computers, or REN and Aveda cosmetics or Chanel, Guccci or Zara fashion brands. They are able to continually develop throughout our lives and like any good relationship can keep our excitement and passion kindled.

DEFINING EFFECTIVE CUSTOMER RELATIONSHIPS

We need to be tuned to the specific part of the customer relationship life stage and direct our communications accordingly. As a woman moves from initial flirting with a brand through to greater engagement and loyalty, our marketing relationship strategy and tactics need to be revised accordingly. Life stage theory, including Levinger's work (1980), has significant value here. With three broad phases (see Figure 8.1), the total lifecycle may take a matter of seconds or minutes for a transient relationship with the newspaper seller or canned drink or last a lifetime with your husband or favorite brand of toothpaste or beer. The greatest insights can be gained from focusing on two areas: the initiation – getting a new customer to try your brand – and the resolution – how a customer relationship breaks down. Action taken at either of these will generate the maximum return on investment (ROI).

Initiation

Every customer friendship begins with awareness and consideration. Just getting onto a woman's consideration list is often the chief task of

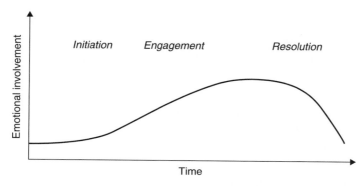

Figure 8.1   Phases in the life cycle of a customer relationship.

marketing. It is also hopefully the optimistic start of a valued relationship for a brand owner. Before all else, the customer must be aware of and engage with the brand; and as with so many other things, first impressions really do count. A "call to action" or a product trial is a key method to engage new customers either to the category or your brand. Women are more persuaded by such offers as they allow them to actually trial a product rather than rely simply on the claims made in advertising. This need for detailed research prior to purchase is a clear difference between men and women.

Similarity of brand values

Research has shown that all customers (women and men) prefer brands with similar attitudes and values to their own. Some will prefer companies that have sound ecological and ethical attitudes that fit with their own. For example, the Body Shop doesn't test its products on animals and this is likely to fit a with a specific customer segment that is against animal testing for cosmetics. Equally, Nike will appeal to alpha females and males that are driven to win – responding to Nike's recent advertisement during the Olympics: "You don't win silver: you lose gold." Similarity between customers' and a brand's values is important for a number of reasons (Byrne 1971). First, customers like brands that reflect their values and beliefs because they reinforce their self-esteem through confirming that their set of values is accurate and correct. Second, brands that have values similar to those of their customers find it easier to communicate with them (and vice versa) and this reinforces the customers' perceptions that the brand is strong, good and right for them. Finally, women psychologically fantasize that because they like a brand, then the brand correspondingly likes them, creating a halo effect on their sense of self.

> The lesson here is that strongly feminine brands are inclusive brands. They build and reinforce social connections, rather than exclusivity.

Women, with their higher social skills, will try and avoid sharp outcast status. Women have a much less divisive split between the most popular and the least popular members of a group than do men, because men are more competitive. This is often witnessed early in our lives through playground groups that symbolically separate the popular from the unpopular. The curve for women's groups is flatter than for men's groups, where there

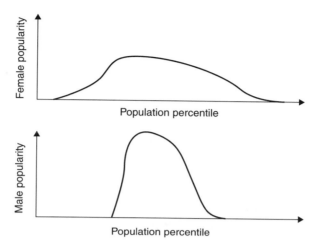

Figure 8.2 Broader (female) and narrower (male) group popularity ranges.

is a more divisive split (see Figure 8.2). We can plot the same curves for gendered brands. Brands that have a more feminine stance within their category – Renault, BP or Orange – have a broader appeal than the more masculine brands like Mercedes, Esso or Vodafone, which will experience a more polarized view of their appeal.

Byrne *et al.* (1986) defined a two-stage process that women go through during brand selection. In the first stage women use repulsion criteria: they de-select brands that are completely out of synch with their value systems.

Many big brands like Exxon, McDonald's and Nike have all come under attack and suffered from brand de-selection, resulting in banishment from a women's consideration sets. Then, in the second stage of brand selection, they assess which of the remaining brands are most similar to themselves and select that brand for purchase. Each of the remaining brands may not match exactly, so women perform a mental conjoint analysis to trade off the least-important criteria. Their greater holistic reasoning abilities allow them to perform this function much better than men.

Dialogue marketing

Women's stronger group orientation means they welcome brands that reach out to them and encourage interaction. But this form of liking is not a simple, direct relationship. We tend to like people more (even more than those that have always liked us) who initially disliked us and then changed to liking us. This is known as the loss–gain paradigm. In simple terms it's about a brand's ability to appear initially aspirational and unattainable or "play hard to get," followed by its availability and accessibility. Fashion brands like Ralph Lauren or Giorgio Armani have built their businesses on this approach. Their initial clothing range is focused on high-end couture. This creates an aspirational image for everyone from high-net-worth individuals to the minimum-wage taxi-driver coveting the products. By portraying themselves initially as unattainable such brands create an unmet desire in women, a psychological void to be filled. As the brand is built and the company develops diffusion ranges such as Emporio Armani, another tier of women is able to finally fulfill their dream. The process of initial unattainability followed by availability offers customers psychological resolution and mental closure. This creates a stronger sense of loyalty than had the product been originally available to all. Following the introduction of Emporio Armani, further brands were created at still lower price points, for example Armani Jeans. These were more widely available and accessible to a wider market, allowing them to fulfill their own dreams and own a piece of the Armani magic. Once again, for these women, the initial rejection followed by engagement creates a stronger relationship with the brand than one that had initially been readily available.

> Portfolio strategy and new product introduction roadmaps are the critical tools to growth of sustainable customer acquisition.

The staging of selective release in waves can be created through a number of different strategies. Giorgio Armani, for example, chose to develop discrete brands with unique brand identities. They all sit well within a family look and therefore support each other. The family of brands differs on three main dimensions. The first is price, with exclusive couture items costing several thousands of pounds. The second dimension is availability, supporting the proximity concept and likability; the Armani jeans are sold through retail outlets rather than Armani's own retail stores such as Emporio Armani or Giorgio Armani. The third dimension is specific to the fashion industry: the brands differ in their level of style formality, so Giorgio Armani le Collezione is very formal while Emporio Armani is a mix of formal and casual, and Armani Jeans is clearly a casual clothing offer. This strategy has been incredibly successful with women for two reasons. First, each successive brand attracts a new set of female customers on an increasingly larger scale from niche to mass market. Second, it encourages cross-selling across the ranges and supports a woman's desire to expand the Armani look across different areas of her life. This results in women buying from specific ranges to mix and match across their lifestyle needs, so they may own a cocktail dress from Le Collezione, a blouse from Emporio Armani and a T-shirt from Armani Jeans.

ENGAGEMENT

We have so far discussed how we can encourage women to form relationships with brands. In order to grow these we need to identify areas of complementarity rather than agreement, something that adds to the relationship rather than simply mirroring it. This provides a benefit to women that increases their overall brand satisfaction. Brands like Zara and Ikea are great examples because they provide women with a sense of style that is far beyond the cloth of the blouse or the wooden furniture. They are similar enough to be considered but also complimentary because they provide the additional emotional benefits. It seems easiest to map out exactly what the level of similarity between customer and brand personality should be; a certain core agreement is required, but real attraction is achieved when it offers something aspirational and additional to the women's personality rather than just simply reinforcing what is already there.

RESOLUTION – LOSING A CUSTOMER

A complete breakdown in customer relations rarely happens suddenly; like in any relationship it often begins deteriorating through conflict or dissatisfaction. If the brand experience fails in some way, then the relationship will begin to deteriorate. Women will begin reassessing the costs and rewards of the current brand against the alternatives available to them. Sometimes dissatisfaction is due to external events that are entirely unexpected, for example a change in public opinion about an issue like smoking or the environment, or in the case of food companies the growing problem of obesity.

Strong dissatisfaction is caused when the actions of the brand negatively affects a woman's self-perceptions. Mercedes recently suffered a dramatic increase in customer dissatisfaction because of quality issues on their cars. Meanwhile their competitor Lexus had increased its quality levels beyond the German car brand and delivered much higher levels of customer service. This meant that customers had active reasons to stop buying Mercedes and active reasons to start purchasing Lexus. Dissatisfiers might be experiencing inferior service at a restaurant, hotel or bank, or inferior or unexpected product performance such as a noisy vacuum cleaner or a TV that breaks down soon after purchase. While it is the specific instance that has caused the dissatisfaction it is the brand relationship that is on trial, not just the product. Very loyal customers will give a brand a second or third chance at recovery after poor performance; but most customers are becoming hyper-critical and recognize that they are in a powerful position and may switch brands easily. Women tend to be more loyal and evenly balanced in their decision-making than men, requiring stronger reasons to switch.

The level of relationship has a direct impact on the level of likely dissatisfaction: transient and superficial relationships rarely have conflict because neither party is very committed. Low-interest products such as household cleaners or lightbulbs will fall into this category. Significant relationships like those with your bank or favorite restaurant are closer and therefore have the potential to result in stronger dissatisfaction as a result of the depth of the relationship. There are three levels of potential dissatisfaction that women experience in their relationships with brands:

Specific brand dissatisfiers – the majority of conflicts

- the food tasted bad
- the service was slow
- product quality was low
- the oil tanker spilt oil on a protected coastline

Dissatisfaction over roles – the role of the brand changes in relation to a woman's needs

- the local gym brand becomes more of a social club than a place to exercise
- your favorite retail brand decides to go more upmarket or downmarket
- your favorite fashion brand becomes an icon for an unpopular segment, for example pro-capitalist/anti-capitalist

Dissatisfaction over values – the brand changes its values, often as the business grows

- the brand decides to drill for oil in a wildlife area
- the brand uses illegal labor practices
- the brand becomes arrogant towards shareholders, colleagues or customers

It is precisely at the point of dissatisfaction that a brand can either lose a customer or gain a new one by encouraging her to switch brand. Strong feminine negotiation and nurturing skills and activities will ensure that the customer does not drop your brand. The key aspect of this is identifying and preventing the dissolution of the customer relationship. One of the best ways to prevent this happening is to understand the pivotal dissatisfiers – things that will make customers walk away from the brand.

> The battleground for customer profitability is during the engagement and resolution phases and marketers need to focus their efforts here.

Reinforcement theory

Companies typically focus on the central elements of the journey as these phases mirror the payment for the service. But women tend to overemphasize the start and end of the experience in their memories, owing to the effects of psychological impact (initial impressions) and recency (the last part of the service they experienced). The initial impression is, like any experience, the point at which women make up their minds about the brand and their preferred relationship with it. Because people make up their minds very quickly (often in the first few seconds) they form a view of whether the service is good or bad and therefore the branded relationship. This means that marketers need to focus strongly on the initial contact and maximize satisfaction of a first trial by customers. Just like making new friends, we make an initial decision very quickly about someone we meet – liking them or not – and then build up reasons why this may be the case. Equally we need to place strong emphasis on the resolution phase. Investing time and energy at these points will shift a one-time transaction into a profitable lifetime customer relationship. All the money spent on acquisition and servicing this female customer will be wasted if they are not converted into a long-term customer.

CREATING HYPERSATISFACTION OR "FLOW"

One of the most insightful descriptions of how and when we feel good is called "flow," a concept originally described by Mihaly Csikszentmihalyi (1990). "Flow state" is when we feel like the world is a brighter place, we are energized and everything falls into place easily and beautifully. We feel like flow when we have just experienced a romantic moment or won a race or just feel like life is going our way. The endorphins kick into our system and spread happy feelings all over our body, promoting mild ecstasy. This can happen when we use a product or service, when everything comes together just at the right moment and in the right way to make us feel really good. We all remember when we have used a service that made us feel immersed, relaxed and blissful. One of us experienced this flying back from India recently. The British Airways Club World service at Mumbai airport is one of the best in the world. From the moment your car arrives at the terminal the BA staff identify you and carry you and your bags past the long queues and the security guard and into the terminal. They give you preferential treatment at the security bag scanner by putting your bags on immediately and then carry them off to the bright marble room that is an oasis of calm in the hectic airport. The whole privileged experience is just how flying should be: stress-free, like floating on a magic carpet through the hordes of travelers, noise, energy and chaos that is modern India. The experience flowed sublimely and smoothly, not disjointedly and abruptly. The seamlessness of interaction, combined with just the right energy levels and appropriate timing, helped to achieve optimal flow. In mature or low-brand-interest categories where there are very few emotional differentiators, the ability to create flow is the key to success. Retail banking is a great example of getting flow right. Since interest rates, availability and range of products are largely similar, it is the actual service experience that really differentiates brands. There is no surprise about what the customer wants: they want their bank to understand their specific needs and tailor their products to satisfy them. While all banks see this as their key goal, it is the ability to deliver against this goal that defines the winners and losers on the high street. Those that get it right are the ones that deliver a "flow experience": effortless, yet engaging (see Figure 8.3).

There are two basic forms of flow. First is the unprescribed or loosely prescribed service, where the customer is often in control of many elements. For example, supermarkets allow you to shop and use self-scanners to register and pay for your goods before leaving, giving you the ultimate

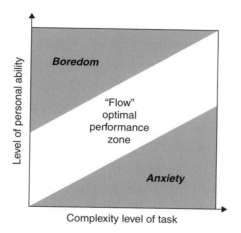

Figure 8.3   "Flow" state optimal performance zone.

control over pace, sequence and so on. The second form of flow is where everything is prescribed but this prescription fits with your needs exactly. The amazon.com website has two primary options, one-click shopping for those in hurry and a more complex option for those with more specific needs. Each of these helps to attune the shopping experience with greater flow for your customers. Flow seems to happen best for women when the experience is a little bit beyond the usual. It needs to be not so different that there is a sense of uncertainty, anxiety or fear, nor so similar that they are bored or the experience requires no effort on their part. This is illustrated by a zone of experience that is a tension between one's increasing abilities and the level of activity one is required to complete. Rather like an elastic band around two fingers, flow occurs when it is tight enough not to fall off, but not so overstretched that it snaps. The optimal tension of the elastic band mirrors the optimal tension that occurs in a flow experience. Optimal levels of flow are accompanied by a feeling of joy, when things are just right.

THE SIX-STEP FEELGOOD FRAMEWORK

The best way to develop flow experiences that make women feel good is to act like a choreographer or conductor: not only to provide the appropriate elements, but also to drive them together at the right time and in the right way. Great experiences flow smoothly. By no means every brand is able to

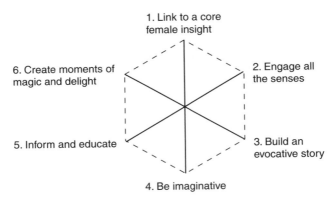

Figure 8.4   Essential components of a feelgood experience to produce "flow".

create the real flow that makes women feel good, because it requires a number of elements to combine effortlessly. The following framework describes the essential components of a feel-good experience to make women feel better about your brand and enjoy flow (see also Figure 8.4).

1. Link to a core female insight

The experience needs to be appealing to women and their specific characteristics, without patronizing them. It is based on the core insights of the target group of women, many of which have been identified in the earlier chapters of this book. This is a crucial part of encouraging women to trial your service for the first time. They may be new to the category – for example they may be using an internet bank for the first time – or they may be a switching customer who is trying a service after being dissatisfied with a previous service provider such as a mobile phone company. Either way the specific appeal should match to a core insight or driver of demand for you target female audience in order to be the most relevant. InterContinental Hotels have developed a new proposition based on a core insight about travelers. Their insight was that in the world of luxury hotels the physical environment of soft beds, power showers and excellent food are expected. What modern travelers want is a flavor of the local culture when they travel. InterContinental's proposition that they are the "In the Know" hotel in every location provides an intangible and emotional benefit that distinguishes them from other hotels. It moves them from being a seller of rooms to one that is selling experiences and this result in higher customer satisfaction and commands a premium price.

2. Engage all the senses

Making a brand more attractive to women customers requires stimulation of all their senses. A beautiful fragrance or a sensual voice can seduce women into trying out a service for the first time. Sensuous appeal uses physical elements to create emotional responses in the woman's mind, encouraging her to try something and convincing her that the risk of failure is low. Hotels like Hyatt and supermarkets use own-branded fragrances pumped into the air to help build a unique experience. Women are better able to see the big picture of an experience and connect that emotionally with the way that they feel. Men tend to perceive the elements but compartmentalize these as more rational characteristics.

Once a woman has been attracted to a brand, she needs to be engaged in a powerful, live narrative. The analogy to theatre is appropriate here, as the woman is located within a space and experiences a service around her. The sense of theatre or drama within a service experience is an opportunity to make women feel great. The distinction of branded services is that they are time-based and therefore the opportunity for live action is a tremendous benefit. To follow the analogy further, service brands can identify their "stage set," namely the actors and the role each of them plays. Women are very used to this kind of multidimensional experience and can easily read the semiotics of retail or service theatre. The fashion brand Zara is expert at creating this theatrical experience within their stores. They offer catwalk fashion styles at affordable prices, but they are able to retain a sense of exclusivity because they have a limited run of each item. This means that the store has a constantly changing look and appeal that keeps customers returning every week. The interior design, music and shop staff all exude a fashionable swagger that lets customers know they are in the right place.

To revitalize its brand among women in the late Nineties, Godiva provided an indulgent experience when it unveiled a new store design, updating its slightly intimidating and austere image with an art nouveau style to communicate a heritage of sophisticated European pleasure. To further transition the brand from a product to a lifestyle into 2002, the company launched new products and packaging, in addition to running a series of emotionally driven print advertisements emphasizing Godiva's products as a much-deserved, self-indulgent treat. Full-page advertisements featured a woman in a museum eating Godiva chocolate above the caption "Who says quality time has to be spent with someone else?" Godiva has extended the campaign playing on its name and positioning women as a "GoDIVA." Godiva's sales have increased worldwide.

> Women respond much more to verbal and non-verbal communications that should be the foundation of all great brand experiences. Be wary of staff that hide behind store props.

Drama creates an emotional connection between the brand and a woman that is crucial to forming strong loyalty with the brand. It is this drama that builds the mental replaying we noted earlier as crucial to forming strong brand memories and loyalty. eBay is a great example of a brand that continues to provide additional drama to the act of buying. Its character of an online market stall is inviting enough. But with the bonus of an element of gaming thrown in it is an unbeatable and vivid brand experience. Many categories that are immature in a marketing sense still believe that functional superiority is a viable approach to growth. The reality is that moving up through Maslow's hierarchy of needs towards more emotional and aspirational characteristics provides a stronger foundation for superior growth and at a lower risk. Of course, drama can be created in many different ways and is extended to different levels depending on the context. Dramatic emotional connections will be very minor for an investment banking experience, although even in this case the energy of the capital markets and dynamic business process of mergers and acquisitions can be harnessed to achieve increased loyalty among female clients.

3. Build an evocative story

"Stories make sense of information." It's easy to see how in a brand experience context there may be an overload of information that needs to be ordered along a timeline for women. But it is how stories make women feel better that defines the appropriate brand experience. Prêt à Manger the sandwich shop (now 33 percent owned by McDonald's) are great at telling stories throughout the customer experience. Their brand positioning is "Passionate About Food" and this exudes from the entire brand experience. They provide instructions on the sandwich pack of how to make "your favorite crayfish and avocado sandwich" or their delicious soups. These all bear the signature of Julian Metcalfe, the founder. Their brand shares knowledge about food rather than just selling sandwiches. They also display high ethical standards by donating leftover sandwiches to the homeless, which also reinforces their "hand made today" promise on all their food. Staff are conversant with every ingredient and share anecdotes about where the ingredients come from and why they are good for you. The story is enriched with physical props like the use of professional kitchen design elements in the store and professional chef uniforms help to differentiate the brand and underline the story. The general manager of each store has a card pledge to serve customers better and deal with the "Good, the Bad and the Ugly" if he can; otherwise he'll forward your comment on to Julian to deal with. It's a classic example of empowered staff who use food as the narrative for a lively brand experience. There are several parts to any great story and we can see how they affect customers within a service context. Each Prêt à Manger service element provides landmarks for a woman to follow. These landmarks are micro marks that ensure women do not stray too far from the desired service experience and that they get the best out of the experience.

Many brand experience designers may believe that once the initial resistance has been overcome and the first tipping point achieved, then women will continue to engage in the story as desired. However, this is largely untrue and very different from how men build branded relationships. Women are constantly reviewing their sense of effort and reward and how they feel throughout the experience. This means that it is not a forgone conclusion that they will both use the service to final resolution and return to the service for a second time without unprompted effort from the brand. This is one of the many characteristics that make women different from men. Men usually have a much higher threshold for change. The brand experience is only as good as the last interaction a woman had, as it only exists in her mind at that last point of interaction. At all other times it exists virtually, but there is very little concrete evidence of this, other than a service contract, welcome letter, weekly update report or other physical totem. Ocado the online grocery brand (in partnership with Waitrose) use a clever stream of communications and offers to ensure their customers stay true to the brand. They carefully track purchase frequency and have analyzed the point when customers start to drift away. They know that if a customer has not purchased in the last three weeks they need to reactivate them with an offer, incentive or news about the latest product or service. This continuous strengthening of the Ocado story helps to remind customers of the brand benefits.

Increase a woman's connectivity

Women invest huge amounts of time and energy connecting with other people because it makes them feel more fulfilled. They love experiences that facilitate these connections, whether they are limited, like a short conversation with the man at the convenience store, or deeper, with support structures in the community, like school governors, social clubs and local supermarkets. The ability to connect with women customers requires employees to have strong empathic abilities. This is not a transactional skill that can be learned but a talent some people have more of than others. Enlightened retail employers have recognized this and have hiring policies that seek to identify personnel who have high emotional intelligence. Charles Dunstone, CEO of the highly successful Carphone Warehouse brand, once said that he tries to hire people who have no previous mobile phone selling experience but who are naturally engaging and like other people. He wanted to avoid their prejudices towards other brands and build a different type of brand culture from scratch. It is the ability to connect with people that he is seeking, rather than a specific sales attitude or skills. As the best employers know, hiring for attitude and training for skills is a more effective route to powerful customer connections and this is especially true for women customers.

4. Be imaginative

Women enjoy time off and relaxing in a fantasy world, one where they can be anyone they want. Women live high-pressured lives and a few minutes or hours off in a world where we can become our role model makes women feel wonderful. The need for imagination links back to early experiences as a child, where most learning was achieved through play. It is only now that businesses and markets have recognized that this is still an incredibly effective way to engage women. The Westin Hotel chain has branded its bed the "Heavenly Bed." This sets the expectation for its customers that they will receive a better night's sleep in this bed than in a normal bed or the bed of their competitors. The reality is that because the expectation was set, then women's perceptions are that they have indeed had a better night's sleep. It is a type of auto direction, encouraging customers to believe the product has had a positive impact on their performance. A powerful brand icon can add a halo effect to the rest of the product range, helping to transfer its power to lesser parts of the range. This is because the benefit of "heavenly" is an emotional benefit and will encourage the customer to feel strongly with your brand. We all have an imagination and like to dream or daydream. But somehow there is an unwritten rule that as adults we are not allowed to spend our time and energy pursuing these benefits. In order to help women release their imaginative talents they need to provide opportunities for escapism, whether this is an urban spa at a Hong Kong hotel, which encourages guests to relax overlooking the harbor while recovering from jet-lag, or perhaps an airline including cartoons in its choice of films so that women and men may watch without feeling "We are not allowed."

Help them relax through games

Games and play are part of their lives and it's often the foundation to making friends. Businesses that help women to make friends with them by engaging them in games and play will find a highly responsive audience. We have only to look at the popularity of the lottery or TV reality games shows to know that everyone enjoys either taking part in or watching games. This can be cerebral, like the intellectual game that the advertisements from *The Economist* engage its readers with – they are witty and subtle; you have to be a member of the club to get them and there is a wry sense of humor about them. Or the games can be more physical, like when cosmetics brands help women with a trial makeover before they buy. eBay is the fastest-growing play retail phenomenon and one of the highest-increasing brands in terms of brand value. According to Interbrand, their total brand value jumped by 18 percent in 2006 to $6.7 billion. The bidding process is effectively a game format that is used by millions of people to purchase goods. The game of chance, about whether you will be the successful bidder and the final purchase price, is something that has proved addictive to many. It is the sense of gaming that is one of the powerful drivers behind eBay's success. Online poker is another huge phenomenon at the moment and women are a well-represented proportion of the gamblers.

Encouraging women to release imagination through your brand experience will increase their emotional connection and therefore loyalty to your brand.

Women love to solve puzzles

One of the skills women are expert at is solving puzzles through pattern recognition, whether this is recognizing the pattern in a musical score or the act of reading – the most basic pattern recognition we each learn as an infant. Women love to recognize patterns for two reasons. First, they help to make sense of their complex world. Second, it provides a sense of mental closure by "completing the picture," giving them the satisfaction of having "worked out the solution." Accessorizing and coordinating are the most effective way brands can help create this sense of solving puzzles. Brands like Ikea tap into this need by providing not just furniture but a whole range of accessories as well. This allows the woman to build her home within a range of elements that will work well together. Exactly how she does this is different for each woman, providing ample opportunity for individual taste and the satisfaction of solving the puzzle of how to create a dramatic room set at home. Retailers like Target and TopShop are also expert at going beyond their basic items to help women customers create solutions from the jigsaw of pieces in store. By leaving parts of the service experience open to interpretation, women gain satisfaction and confidence from mentally completing the story. When a brand is prescribed down to the finest detail there is nothing to be left to the woman's imagination. Her capacity for comprehending the big picture is far superior to a man's. Working out how to get the most out of a service can increase a woman's sense of mastery and self-worth. She starts to feel like a savvy expert. This sense of getting better at something is an important human aspiration.

Help them fantasize

Women, just like men, like to imagine themselves as more beautiful, sexy, friendly and funny. Brands that make them feel special are always going to be successful. Businesses should leverage this opportunity by using brand experiences that enhance women's sense of fantasy and escape. For an online bank this may work by treating a woman as though she was a millionaires. Even if she only had a small credit or debit account, she could still be made to feel like a VIP every time she spoke to the bank. It wouldn't cost the bank any more money to change the way they spoke to women but would have a huge impact on their perception. Virgin Atlantic is a great example of a brand that encourages women to fantasize. Their party-in-the-sky attitude means that every customer feels like a high-flyer. Even naming their business class "Upper Class" demonstrates Virgin's commitment to making people feel out of this world. Their range of onboard services including a massage and excellent goodie bags for children helps reinforce this image. Virgin has an open bar area on the upper deck that reinvents the flying experience of the golden age when only pop stars and royalty traveled this way. It gives everyday travelers a slice of the jet set lifestyle.

5. Inform and educate

The flip side of indulging fantasy and escape is for a service to engage women through a sense of growth and learning. This can be provided on two levels: the learning of facts and knowledge, or the training in a certain skill. Brand experiences can help women become more informed by gaining knowledge and facts. InterContinental hotels' "insider experiences" are designed to provide just that level of entertaining education. They offer customers options for insider leisure, culture, occasions and shopping. For example one of the insider culture offers is a private wine tasting at their hotel in Wellington, New Zealand. Guests can stay at the hotel and gain insider knowledge about the region's wines while enjoying them and the gourmet food. There are many other businesses, especially in financial services (both retail and investment banking), where knowledge about the best deal is key to power. The old stock markets were based on advisors predicting which stocks would rise the most. The additional benefit of feeling more financially astute through using these services feels tangible but is hard to quantify.

Because the nature of brand experiences means that they are often delivered by people, it is inherent that those people are expert, or at least more expert than the women they serve. This means that it is relatively easy to impart knowledge to women during the actual experience. The challenge for brands is to recognize that the amount of knowledge required and the way it is expressed are critical in satisfying women. For example, the reason a woman is using a service is very often because she has insufficient knowledge. She may therefore very well wish to remain relatively ignorant about this topic. However, if the relevant information is provided in handy, bite-sized chunks that are easily digestible, it is more likely that women will appreciate the additional benefit of that knowledge. In a restaurant, learning which wine goes better with the monkfish is another by-product of the intended service. There is an analogy here with Karl Popper's famous philosophical declaration: "There are always unintended consequences of intended actions." Translated for a brand, this means that there should always be additional benefits for women above and beyond those prescribed in the service contract. Very often these additional benefits are in the form of knowledge or learning that make women feel good. The tone this kind of education in is delivered is also critical. Using a suggestive approach is always more effective for women, as it not only imparts knowledge but also makes them feel important. For example, a waiter may see a woman glancing over the wine menu because she is unsure about which wine to choose. The waiter could simply ask her what kind of wine she is looking for, but this might embarrass her in front of her guests (assuming she knows little about wine). Or he could make the woman feel good by using a suggestive approach towards them: "I'm sure you were considering either the Chablis or the Sancerre with the sea bass."

> Women appreciate practical advice and gain self-confidence through mastering new ideas and services.

Training is another form of gaining knowledge, just of a more practical kind. Many brand experiences can be explicitly designed to help train women. These can range from a personal trainer, who helps get them into shape, motivates them and advises on the best diet and training regime to achieve their goals. The same is true of language training or piano lessons; the brand-provider explicitly trains the customer in a new skill area. There are other brand experiences where the training is implicit in or additional to the primary purpose of the service. These could be a restaurant that implicitly shows you the best way to eat sushi with chopsticks, or the best cut of meat to choose from the menu. Training and learning is an inherent desire for humans; they largely enjoy the sense of continual learning as they easily get bored with repetitive tasks and routines. Combining some form of implicit or explicit training and learning within the brand experience will encourage women to remain loyal and become great advocates for the brand.

6. Create moments of magic and delight

Powerful brand experiences are those that have personal significance, making them unforgettable for women. The experience has to encourage or enable a shift in their attitudes, beliefs or behaviors; it should make them a different and ideally a better person. Visiting the Grand Canyon changes

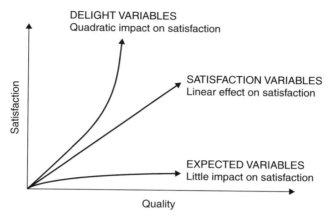

Figure 8.5 Exponential increase in customer satisfaction from delight variables.

a person's perspective on nature, in the same way that visiting Dubai or Las Vegas changes a person's perspective on luxury hotels. These feelings are generally those that move beyond the functional benefit and deliver a truly emotional experience. A recent statistical analysis of ideal customer experiences highlighted that satisfaction variables can significantly increase the return on investments made specifically in touchpoints that deliver "delight" to a customer (see Figure 8.5).

> Random acts of kindness generate huge word-of-mouth power with the sisterhood.

Even in the most mundane of services, there is room to surprise and delight women through one-off actions. These may be as little as a thoughtful touch, where an employee recognizes a long-time female customer and rewards them with a free cup of coffee, or an upgrade to their service. It may be where common sense prevails and the employee delivers something outside the normal service parameters, like a hotel butler going out of the hotel to buy a female guest's favorite toothpaste. In order to achieve memorable brand experience delights, employees need to be empowered to recognize that the service standards do not always have the appropriate solution for every woman. The size of tolerance for this kind of activity needs either to be defined by the department manager or else codified in some way. A luxury hotel chain allows its employees the power to solve guests' problems up to a limit of $2000 US. This can make a female guest feel like a VIP, by instantly solving her problem and not quibbling about the additional cost. The challenge for businesses is to recognize that savvy women and men may take advantage of more junior employees to receive this kind of behavior. It is open to more abuse not only by guests but also friends of employees who can be over-compensated by insider connections. The best service delights are those simple additions that show that a firm has truly understood women's needs.

Women are more open about consciously indulging themselves than men. Indulgence may be seen as not goal- or task-oriented and therefore as an area in which men are less interested and less able to demonstrate their performance. If men turned their minds to it, they could be good at indulgence, but only in a solution-driven way, which rather defeats the purpose of indulgence. In this case indulgence is about pampering oneself – investing time in oneself – to feel happy. Women tend to use shopping, wellbeing, food and drink as ideal indulgences. Men prefer technology gadgets and sports as their indulgences. Brands need to position themselves as offering women a sense of indulgence. This might be the benefit of extra time so that they can indulge in their favorite pastime.

Making women feel indulged is a result of generosity shown throughout the brand experience. This means that employees need to be generous with their time, and pace themselves at the same speed as their customers. First Direct bank has a rule that their call center employees should always take as much time as each customer needs, whether they have called to make a brief check on their bank balance or to have a lengthy and engaging discussion about their banking facilities. Either way, the call center employees are generous and indulge their customers for as long as they like, rather than competing to complete as many calls as possible. It shifts the focus of their brand experience from quantity to quality, ensuring that their customer feels good. Generosity can also be demonstrated through the attention that employees give to women customers. As we have discussed earlier, women like to feel they are being listened to, even if they are not expecting an immediate solution. For women, talking through the options in detail may be all they require from retail assistants. Attention can come in many forms; women like it when people show understanding of their situation and are generous with their ability to adapt to their individual circumstances. A harried mother with two small children will appreciate it when the sales clerk produces a couple of toys for them to play with while mother chooses a new pair of shoes. This kind of empathic and generous attitude is rarely displayed by goal-oriented men; but it ensures that the brand is unforgettable.

SUMMARY

Women build relationships with brands in the same way they make friends in the neighborhood. They have a range of different friendships with brands from intimate to transient. Marketers need to map out these relationships to understand how to increase the potency of their branded relationship with them.

Branded relationships follow the same life cycle as human relationships, from initiation through engagement to final resolution. Mapping this out against female segments and their needs will identify actions to improve customer acquisition and retention.

How women feel is more important than what they achieve, whereas men by contrast focus on tasks and outcomes before feelings. The concept of marketing buzz describes the human characteristic of "flow," the basis of feeling good. This comes about when someone feels at the top of their game – their abilities match the task but with a little stretch. They are neither bored (under-challenged) nor stressed (over-challenged) and everything falls into place beautifully. Endorphins spread throughout our body, creating a tingling feeling of mild ecstasy. Successful brand experiences are able to make women feel fantastic, and this builds stronger loyalty to the brand.

The six components of the feelgood framework help marketers to redefine the brand experience and change the way women feel about the brand and therefore their relationship with it. To be most effective these components should be used holistically as an integrated marketing approach towards women.

9

Touchpoint improvement

This chapter identifies how marketers can use a woman's talent for making friends as the foundation for building potent branded customer relationships. The specific ways women make friends provides powerful insights, strategies and tactics into how to improve these customer relationships. By looking at the life cycle of relationships it is easy to identify critical relationship-building and -destroying touchpoints along their customer journey. This chapter illustrates in detail the business process of improving high-impact touchpoints with women. This includes a detailed brand-driven process reengineering tool that aligns women's needs with improved branded moments of truth. This is a proven tool that has been used successfully with many leading brands around the world. It specifies techniques to help identify core needs, to improve service design and to develop the business case and the return on investment (ROI) case for marketing to women.

OBJECTIVES

- Demonstrate leading practice in marketing to women across key touchpoints
- Describe the brand-driven touchpoint improvement tool that analyses and improves the way a brand connects with the target female audience
- Build the business case for marketing-driven change that will impact women's lifetime customer values

BEST PRACTICES IN MARKETING TOUCHPOINTS

Online

Research figures from Verdict Research show that women spent online an average of £579 during 2005 while men only spent £541 (quoted in Hargrave 2006). And a US survey found that 66 percent of all money spent on the internet is spent by women. There are now more women using the internet (in the US) than men (emarketer 2007).This may seem surprising for a tool that began its life for young, geeky men, but the modern reality is that women are at least equal to men in their use of the internet and have different needs and ways of navigating online. The challenge for website designers is that most of them are men, and most sites have been designed around other men without a thought for what women want. Carly Fiorina, Hewlett- Packard's ex-CEO had a vision of the future of the internet, wanting to make it into "something personal, intimate, warm, friendly and most of all useful." This sounds more like the feminine ideal than its current masculine cold, technologically focused, product-satiation-driven state. Kate Burns (2005) MD of Google UK's AdSales, highlighted that women in the US have been outnumbering men as the major internet users since 2000.

A few facts about women online:

- 75 percent of women shop online to save time and avoid queues, while 63 percent of men shop online mainly to save money (auroravoice.com 2004)
- 40 percent of online gamers are women (Burns 2005)
- 51 percent of women (versus 41 percent of men) are enticed by free shipping during online purchases (auroravoice.com 2004)
- 19.7 million internet users in the UK are women (Nielsen 2005)

Women are more likely to provide personal information as they sign up to a website, since they view this as building an ongoing relationship. However, this will only be done once a base level of trust is established (Schaffner 2000). While most Westernized countries have high internet penetration of around 70 to 80 percent, emerging countries are catching up rapidly. Chinese internet usage is growing strongly, with 137 million users or about 10 percent of the total population. Those connected spend on average 16 hours a week surfing the web, a much higher percentage than in India which has a higher proportion of young people. Internet penetration in India is 40 million or about 3.5 percent (Internetworldstats 2007). According to John Rodenburg's (2003) global internet study both usage and e-commerce are increasing dramatically, and while the US still tops the list of both usage and conversion, many other countries are catching up rapidly. Scandinavian countries like Denmark and Norway as well as Hong Kong, Australia and the UK have both high usage and conversion rates. Generally there is a direct correlation between total usage and total conversion rates. However, Japan is one country that has extremely high internet usage by women but also a very low e-commerce value. According to Onya-san (2006) of ANA Hotels Group in Japan this is because the Japanese are a media-saturated country on the one hand and still very conservative on the other. Japanese women still require a high degree of trust before they will use their credit cards online. Many women shoppers still prefer to undertake research on the net and then visit the local store for the actual purchase. Similarly, online purchasing in India is being driven by low-cost travel brands like SpiceJet Airlines and Jet Airways. Their low-cost business model relies on them using direct channels rather than expensive neighborhood travel agents and this means travelers have no option but to pay for their trip over the internet.

> Women maximize the potential of online and mobile media because it provides them with intimacy and helps them further their social networks.

One the reasons why online shopping is so suited to women is that the internet helps women to get more done in less time. They are able to research purchases in detail, and buy and organize delivery in a few minutes and at a convenient time for themselves. Women and men shop differently; women see it as part socializing, part buying and part browsing and researching. Web shopping must mirror these functions by providing more browsing, chat room and community capabilities. Lands End, the clothing brand, changes its site for women more frequently than the men's site to acknowledge this. They also provide an online chat capability, call-back phone calls, four-hour email response times and an online virtual model that can be modified to fit the female customer's shape (Ragone 2000).

Generation Y women are the most internet-savvy, with 83 percent of them using the web regularly (Armstrong 2006). The teen years are a crucial part of the lives of these women and the brands that hope to build long-lasting relationships with them. They are in the process of consolidating their sense of self-identity and are confident at choosing brands that help shape and define them as individuals. Websites like handbag.com in the UK or ivillage.com in the US are leading beacons for women because they are designed in a format more attractive to women. They are more like online magazines and cover a range of issues from shopping through horoscopes to love and sex advice. This kind of multifunctional portal is too confusing for men who would prefer to go to specific expert sites for each topic area. For women, this kind of multi-topic site reflects better the way they view life: as an integrated whole rather than the compartmentalized view men hold: "Women like sites that look good and speak to them more about their life" (Cairns 2006). However, when navigating a specific site women are more single-minded and task-oriented than men, who tend to surf more experimentally than women, hoping to stumble across what they want.

Contextualize their life

Boots ran an online campaign that demonstrates what women want: a brand that understands their lives. Boots uncovered an insight that women are often trying to tackle many issues at once – like cutting down on drinking or smoking, while dieting and undertaking a new gym regime. They responded with the "Looking great is easier of you just change one thing" campaign, emphasizing that it's okay to deal with one issue at a time. It demonstrates a deep understanding of the self-consciousness that some women feel about the way they look and encourages them to believe that there is little they need to do. According to Neil Eatson (2006), of Zed, New Media, women are less persuaded by direct-selling techniques that would be appropriate for men. This means that pop-up adverts will really work only when they connect clearly to the actual web page a woman is clicking through. Though men may be diverted by online poker or car adverts while they are surfing for sports news, women prefer advertisements when they remain within the context of their current frame.

By 2010, women are expected to control $12 trillion, or 60 percent of America's wealth (businessweek.com). Citibank recognize this fact and have launched an online membership program specifically targeting women (citibank.com/womenandco). Acknowledging that women have little specific financial advice or products, their mission statement explains:

> Women & Co. was created to address the unique needs of women as they seek more command over their personal finances. The goal? To help you feel that your money is working for you—not the other way around. (Women & Co. mission statement, citi.com)

Women & Co. is an excellent online resource center that offers education about financial products, online advice, tools to help manage finances more easily, and a network of professionals to help make decisions. Citibank cleverly and insightfully contextualize this need for greater financial control by recognizing that women in particular undergo more life-changing transitions than men. These obviously include having children, being a mother, managing the family's finances and, as women typically living longer than their partners, becoming a widow with often many years of their life left to live.

Build editorial content

Rodenburg's study (2003) pointed out that women in most of the major countries around the world are twice as interested in gaining information as men are when using the internet. The next step beyond simply contextualizing a brand within a woman's life is to build or use an editorial portal around specific content areas that attract women and then use subtle adverts to offer them opportunities to purchase. Sony used the popular women's portal shinyshiny.tv, "A Girl's Guide to Gadgets," to advertise its new high-definition televisions and cameras directly to women rather than the sony.com website. The site has blogs, downloads and information on a variety of topics that provide editorial content and advice for women as well as factual information. Other portals like ivillage.com and dailycandy.com have used rich content to build an emotional connection with specific groups of women. Such sites have editorial sections that are sponsored by big brands like Gillette and Coca-Cola rather than just straightforward adverts pushing their product benefits.

One third more women than men are looking for sites that select information for them personally rather than simply targeting the general public. This indicates a desire for emotional involvement that micro sites can provide far better than can global corporate sites. While women-focused portals begin to offer a context for women consumers, brands can retain stronger control by creating micro sites that offer rich, emotionally connective online experiences. The dove.co.uk site is a great example of an informational site that educates, informs and inspires women to be themselves and be beautiful. It's a soft-sell site that is on the woman's side, rather than the hard-sell, immediate-gratification sites men prefer. Another site is givesimple.com that effectively combines great products and ideas with strong editorial content and a blog under its "Livesimple" chapter. This provides women with information, support and ideas about living a simpler and better life.

> Women appreciate the magazine style and editorial content of websites more than men.

While offline magazines are used by women primarily for relaxation, online magazines are used more for their educational and informational benefits. Offline magazines like *Cosmopolitan* tend to have more than a third of their pages as adverts and these also act as fashion guidance and inspiration. However, online, this kind of overt selling is less welcome and women want more editorial content. This might be tips and advice on anything from their relationships to their health. The online sites also provide an instant dialogue on the hot fashion and style issues of the day. Like many sites they are able to run continuous blogs that act as virtual chats that might normally take place over a cup of coffee. This exchange of experiences is a crucially different facet of online magazines and women rate this benefit extremely highly.

Make it easy

Women generally spend more time and effort researching their purchases, whether offline or online, than men. This means that marketers need to make their websites very easy for them to navigate. Rodenburg's study (2003) indicated that women find practical advantages like ease of use 30 percent more important than do their male counterparts. Women find that a company's brand site can often be too technical in layout or detail. They prefer websites that make it easy for them by translating the facts into a language they understand and in a friendlier tone. When it comes to building a website, it is therefore crucial to create pathways that women can easily return along for further investigation or confirmation. They have to be able to hook straight back into where they left off. It's a bit like an ongoing game of chess or a conversation. The thread is still in a woman's mind, but she needs to both review previous thoughts and then move forward another step. This step-by-step method of online purchasing allows women to be comfortable moving at their own pace. It reduces the risks for them and increases the likelihood that once they do finally purchase they will be very comfortable with the brand and become a long-term ally. This latent conversion model requires more patience from online retailers, who previously may have expected their male shoppers to simply log on and hand over their credit card details. However, this male speediness also comes with a higher level of retail promiscuity. The smarter brands are helping women to move along the journey at their own pace, knowing that their investment in these women will be rewarded with greater long-term benefits.

Advertising

Advertising is a communications channel that needs to respond to the trend in feminization and change its portrayal of women more than other channels. It is particularly prone to stereotyping women and men, but the women are typically depicted more negatively than the men. In a recent study 50 percent of women said that advertisers portrayed an old-fashioned view of women and an unbelievable 91 percent of women felt that advertisers did not understand them (Armstrong 2006). Women are still too often illustrated as an object of male desire. This approach tends to fall into a few recognizable characters. There is the youthful nymph, suggesting virginal purity, innocence and gentleness. Second there is the siren, a femme fatale or free-willed temptress who exudes desire, mystery and sexiness. Third there is the mother figure who connotes caring, homemaking, nesting, nurturing and wellbeing. Finally there is the matron or grandmother figure who communicates worldly kindness, serenity and patience and offers selfless encouragement. This is a narrow set of characters for advertising creatives to draw on given the distinctions highlighted between Generation Y, Generation X and Baby Boomer women discussed earlier.

It is also notable that the range of potential male characters used in advertising is at least double that of women. This seems likely to reflect the proportion of men working in advertising and the greater maturity of the practice of advertising to men. It was probably only forty years ago that women began to be an advertising target in their own right. Early stereotypical advertisements focused on the good mother character, suggesting

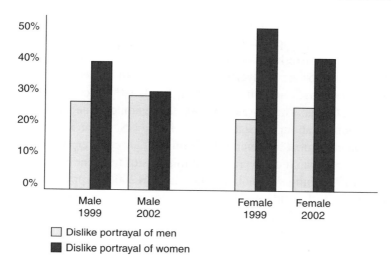

Figure 9.1 Perception by gender of the portrayal of men and women on screen (Source: Salmon 2005).

that without the new Brand X stove she could not be complete as a woman, nor be a good mother to her offspring. Recent research by nVision (2003) suggests that women are far less happy with their portrayal on screen than men (see Figure 9.1).

Advertisements today still use women in the traditional roles of housewife, seductress and mother. Chocolate is sold exclusively on the basis of soft-focus seductress imagery, even with simulated erotic acts taking place with chocolate bars. The mother stereotype often depicts the woman at the center of household chaos. She is the figure in the center, like a spider controlling and managing the entire family. The role of central pillar of the family is still the reality for most mothers, working or otherwise. Some advertisements do show women in a business context but they tend to be masculine versions, women who have resorted to conforming to the masculine rules in order to play in the business world, rather than as women in their own right. The challenge with a lot of these advertisements is that none of them actually show real women. Women viewers therefore fail to recognize themselves within these advertisements and get turned off. While sometimes they may appreciate the aspirational image being offered, women are too savvy to not see the selling message that accompanies this. The point is that in order to overcome these stereotypes, advertisements need to demonstrate a greater level of insight into the female psyche. Advertisers need to target women by appealing to their social strength. They must also recognize that women use television advertisements to remind them or for fantasy and escapism. They get their real advice on which shampoo to buy from their sisterhood network, real friends they can trust. Television is still the most powerful advertising medium, with 98 percent penetration in Western countries and similarly high levels (91 percent) in emerging markets like China. There are over 3000 television channels spread across that vast country, about twice as many as in the US. Chinese Generation Y viewers watch almost as much television as their counterparts in the US. Young Chinese watch on average 162 minutes of television per day compared with 171 minutes in the US (timeasia.com 2006). Apple's advertising has always been legendary. The new Apple iMac and iPod's advertising evidences the feminine, humanized DNA of the brand – "Say hello to iPod" – that is engagingly simple. It uses not the masculine over-selling approach, but a refreshingly simple, emotional appeal to consumers. Women appreciate this simplicity and honesty; the advertisement might not be enough to tell them everything they need to know before making a purchase, but its simple presentation creates an appealing starting point for them to consider the product.

Face to face

According to the stereotype, women are better at conversation than men. However, this is based not on a shallow idea of gossip, but on fundamental neurological differences between the sexes, as mentioned earlier. Women's brains have a specific area in the left hemisphere that deals with language and communications, while neuroimaging maps of the male brain have shown that their brains have a more dispersed activity across the entire left hemisphere. This suggests that women have a more developed and focused area of the brain, the front left hemisphere, that deals with speech. This skill is already developed at an early age. Studies have shown that at three years old girls have a vocabulary twice the size that of boys. Women have another superior ability over men: they can talk and listen at the same time (Pease 2001). This allows their conversation to be more free-flowing, switching between subjects if necessary. It is a form of multitasking that keeps the pace of the conversation higher than that of men. Marketers should take advantage of this by communicating more with women, and using more sophisticated communications means to do so.

One of the key reasons for this is that the two hemispheres of a woman's brains are much better linked than a man's. The left side with its strong linguistic capabilities is more directly wired to the right side of the brain with its emphasis on emotional cognizance and management. This is why, and everybody intuitively understands this, women are better at talking about their feelings. This creates a huge difference. Not only can women converse better; they can do so on a higher emotional level. In contrast, men's communications are relatively simple, factual and stumbling. With this in mind, it's easy to see why women marketers are able to build stronger marketing communications than men. They are better able to use emotional cues and messages that will align with women's desires. As marketing becomes ever more emotive, customer communications require more refined language and content. Given the huge disparity between women and men across all of these issues, it's easy to see why feminizing marketing will increase performance significantly. The emphasis on long-term relationships directs effort away from the transactional to the collaborative.

Word-of-mouth marketing

According to Nicola Armstrong (2006) from Iris Female, "70% of women learn about a new product from another woman," and that makes women the masters of word-of-mouth marketing. Many more women rely on word-of-mouth marketing than men. While men will hear advice, they believe it's important to make their own mind up about a store or product brand. Women, however, tend to shop more on recommendation. These are two things that word-of-mouth marketing can affect enormously. Women will give strong credence to a friend's recommendations about which stores and products to trust. This personal recommendation will often be twice as valuable as a recommendation in a magazine or newspaper article. Any business that wants to sell successfully to women needs to develop their word-of-mouth marketing program. The most important way a brand can increase its reputation with a woman is to get talked about by her girlfriends. As we have seen earlier, women's desire, ability and satisfaction in building strong social networks is both a barrier and a bonus when marketing to them. They will talk about your brand at any point in the day or week. However, the stories women tell can be either good or bad, so brands need to manage their reputation closely. They do this in an intense way that men can only imagine and struggle to understand. A senior woman executive of a large supermarket corporation told one of us how she was in the CEO's office one day chatting to his PA about what kind of shampoo they both used. On seeing the female executive, the CEO came out from behind his desk to join them in conversation. However, something strange yet familiar happened. The CEO quickly realized this was not a conversation that he wanted to take part in, and the women felt uncomfortable about sharing their inner beauty secrets with another man. The story highlights that difference between women and men. Women get most of the brand guidance via the female grapevine, while men would rarely ask or give advice on the shampoo they use to another man. Even for typical boys' toys like televisions or hi-fi's men will rarely offer, request or share their opinion with other men. They like to believe they can make their own mind up and rely a lot less on the advice of their friends. Women for all their new-found liberty still enjoy and heavily rely on the opinion of their friends to help choose brands; in the words of Emma Laney (2005), "Women have gained independence from everything but each other."

> Building viral, word-of-mouth networks is the most cost-effective way to market to women.

For women, there are certain categories that are more likely to be word-of-mouth conversations than others. These typically fall into four groups. The first group is exciting products like DVDs, CDs, clothes or cosmetics that easily become central topics of women's conversations. The second group is service products like hair salons, airlines, vacations and banks where personal experience can help them avoid mistakes or problems. The third group is complex products, where sharing information with friends helps to increase understanding and reduce the risks. These could be mortgages and other financial services products, or computers or mobile phone tariffs which are difficult to fathom. The final group is where expense is the key issue; buying a house, furniture, weddings or vacations are all examples of highly discussed topics. "Regardless of her age, or how busy she is, she still craves time to hang out with her friends. Make your message easy to share and harness the power of the girlfriend grapevine" (Rethink Pink 2005). Some women are better networked than others and will act as much better conductors of your brand story. Depending on the type of product or service, different types of women and social networks will prove more effective. In a recent Indian FMCG brand study, a group of "Mindspace" consumers were identified as the emerging influencers within urban India. These women were 17 to 28 years old, part of the new, financially enabled middle class with a nascent materialistic attitude. They wore Western-style jeans and blouses and were adept mobile phone texting (IMRB 2007). Procter & Gamble also has a huge consumer panel of several thousand participants in the US called "Consumer Corner." They get advanced information about new products and brands, trial these and give feedback to the company (joincc.com). Not only is this panel influential in fine-tuning brand development; they can act as initial viral marketers for the company by talking about the new products that they appreciate with their friends. Marketers can benefit hugely by tapping into these kinds of advanced consumer networks.

Packaging

A quick look at any supermarket's shelves might suggest that pink packaging is all that a brand needs to attract a female buyer. Not surprisingly, the female population is disenchanted with this continuing stereotype. There is no other major marketing audience that is dumbed down quite so far as a single color with no nuances, no variations, just pink. The reality is that good packaging design for women is very similar to good packaging design full stop. It needs to attract a specific audience by reflecting their life stage needs and desires, so what works for Generation Y women will work less well for Generation X women. The pack design needs to demonstrate the benefit to a woman's life, rather than just emphasize a list of features. This might be through focusing on desires like the convenience it will give her, the freshness that will provide her with a healthy life or the beauty it will deliver to her soul. Baby Boomer women don't require superfluous packaging, but they do want to be able to read the type on the back of the pack and open it easily without the help of their partner. It is these simple, inclusive design features that are well received by women, not the explicitly overt women's packaging like pink shades and ribbon wrapping.

Wine brands like Jacob's Creek, Ernest & Gallo or Hardys have all used their packaging design and marketing efforts to benefit from and drive increasing wine consumption by women. A recent Gallup survey in the US highlighted that 55 percent of wine is now bought by women (Feirling 2006). First, the brands here have redesigned their labels to look less like a Latin lesson and more like a welcoming lifestyle brand. According to the survey, women are more drawn to scenic and modern lifestyle visuals, while men remain impressed by images of old chateaux and heraldry. Second, they eschew the masculine sommelier knowledge of varietals for the consumer-friendly world of brands. Women are also more likely to ask for advice in the wine store than men. Women are now more aware of the brand of wine first, followed by the specific varietal. Finally, the brands have targeted women with advertising in conjunction with female-friendly television; Jacob's Creek used the title sponsorship of the *Friends* television series to connect directly with female audiences. This has been a huge success for the wine marketers. In the US, Californian vineyards have introduced "women-friendly" wine brands including "White Lie" and "Working Girl White." They are distinctive and speak in a brand language women prefer. These brands have innovated and contemporized wine packaging. For example they have hidden descriptions on the cork of White Lies wine bottles and a gauze fabric that wraps a bottle more like an exquisite present. Neither of these things would appeal particularly to men, who prefer their wines old-school and complex.

Product

Gillette spent a fortune developing the Venus razor to offer women a new way to shave their bodies. They redesigned the handle, the blade and the lubrication strip to create a razor with huge functional benefits specifically for the way women hold and use a razor. When they launched it in the marketplace they deliberately chose a turquoise color to avoid stereotypes. "We really wanted something that women wouldn't think offensive," says Michelle Szynal, head of global communications at Gillette (Korn 2005). Once this had proved hugely successful they began to offer the product in a range of colors including a pink version. This proves that it is not that using pink should be off limits for marketers, but that they need to focus their efforts on what is important to women. Launching a microwave oven in pink is just incredibly lazy marketing that deserves to be snubbed by women.

In marketing sports-utility vehicles, GM utilized a unique way to connect with its target female audience. It offered $100 spa certificates to one million women for test-driving a GMC vehicle. This effort largely targeted professional women – GM's core audience. Campaign significantly boosted incremental sales growth and drew new female customer to the brand. In addition to the campaign, the company generated brand visibility among a larger segment of women by sponsoring the Women's National Basketball Association (WNBA). This partnership enables GM to leverage the league's extensive media assets through an integrated marketing approach, including in-game advertising, courtside signage, and an interactive presence on the WNBA's website.

The Volvo brand recognized earlier than most car manufacturers that women are important decision-makers and users of their product. It was fifteen years ago that they set up their Female Customers Reference Group to ensure that their needs were effectively incorporated into new car designs. In the US, Volvo cars' biggest market, women make up more than 65 percent of their customers (Kirk *et al.* 2006). In 2003 Volvo's car division reviewed their current global research to discover what were the differences in needs between women and men for three levels of vehicle – premium, mainstream and budget – and found varying degrees at the different levels. At the top level, women wanted the same premium design features like climate control and leather upholstery as men did. But they also wanted more features than men and were more demanding than men in terms of expectations of premium-level features. These additional features included good overall visibility, good maneuverability, easy parking, ease of getting out of the car and convenient controls. At the mainstream level there was little difference between the needs of women and men in terms of their vehicle needs. But at the budget level women expected much less than their male counterparts. They had significantly lower needs on engine performance, seat comfort, the latest safety features and passenger comfort levels. This reinforces the perception that women are not a single group of like-minded people. There are clear differences in needs between different types of women as well as consistent differences between women and men. Marketers need not only to understand the differences between women and men, but also to avoid treating women as a homogenous group of customers or assuming that their differences will be the same right across their product range. Greater segmentation of female audiences and deeper insights into their nuances will help marketers avoid stereotyping their customers.

   Lex Kerssemakers, head of Volvo cars product planning, managed the development of a new concept car by an all-women team (Tweedale 2004). He

> [r]ecognises the commercial benefit of involving women at every stage of car manufacture. Women are more critical and have far higher expectations than men he says. It's amazing that nobody has done this before; we are failing over half our customers by not considering their needs.

although, as Anna Rosen, leader of the team, asserted, "This is not a women's car but a car designed by women for everyone." The resulting car design was one that met the expectations of women and exceeded those of men. It introduced many new features that had for years been part of car tradition: the windshield wash filler and gasoline filler were part of the same opening, rather than having to open the hood each time the driver needed to fill up the windshield washer. Small features, but they demonstrate that starting with a woman's needs in mind can expose long-held and flawed design details. The car was voted best concept car at the Geneva Motor Show in 2004 (Kirk *et al.* 2006).

High-impact touchpoint improvement process

The following is a six-step guide to building more powerful customer relationships with women. The basic approach is to understand the current situation in terms of service and expectations. Define the ideal state from a woman's perspective (some women's and men's expectations may be totally unrealistic). Then redefine the future situation in detail, defining key processes, roles and behaviors. Change is always difficult and a balance between accepting the old organization and creating a new one. This requires a matrix of skills from analytical and formal together with creative and informal. We must thank Tom Agan, a specialist management consultant, for his excellent work on brand operationalization. It was Tom who first kindled our interest in the possibilities of brand service design and the following process has been redeveloped from our initial conversations. This six-step virtuous cycle leads to world-class customer relationships that rely heavily on feminine attitudinal and behavioral traits. The importance of continuous refinement cannot be underestimated in any business that relies on its live performance to satisfy.

Identify key consumer insights

As we have seen earlier in this book, it is almost impossible to design great relationships with women without clear knowledge of their specific needs. This means going beyond traditional demographics and identifying attitudes and the wider lifestyle of the audience. Even for businesses that have a broad offer, it is paramount to subsegment customers in order to gain further insight into their needs. Once this has been completed it is possible to move through the following cycle and build powerful relationships with them (see also Figure 9.2).

1. Assess current touchpoint experience

The first step to developing any customer relationship with women is to understand what is actually happening today. This requires an objective

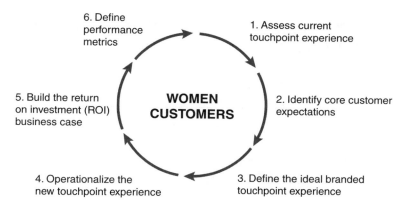

Figure 9.2 Stages in building a powerful relationship with women customers.

assessment that maps the current reality that women experience rather than what should be happening. Marketers need to guard against pretending that everyone follows standard operating procedures (SOPs). Observational techniques such as shadowing or ethnographic research can help uncover these realities, as they identify real practices rather than those that have been prescribed. People often know the "correct" way to do something but many often choose to ignore these rules to save time or energy – they are human after all. A fine balance needs to be struck between focus and comprehensive inclusion. Using a woman's perspective is always the best way to help define and separate those parts of the service that are connected in their minds. A small piece of exploratory research may help in this process (see Figure 9.3).

Identify ideal customer experience

The brand journey women experience is an ideal route map to navigating core tasks and activities. It is worthwhile extending this sense of the customer's journey prior to and beyond the actual service to ensure that the relationship-building initiative has a clear context. This may include a handover from other services or service providers, such as arriving at an airport by a taxi service or train before joining the airline service that propels you along your flight journey. By understanding the wider service context it may be possible to combine, extend or reduce the total customer journey and your involvement in it. As we have seen earlier, the customer

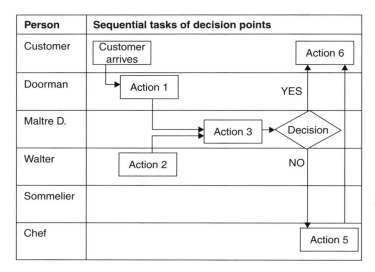

Figure 9.3 Mapping customer service interactions, tasks and decision points.

journey is typically broken down into three phases: initiation, engagement and resolution. We need to consider both the functional and the emotional experiences that occur across these three phases.

If raising awareness with a new female segment is the biggest issue, then clearly the initiation phase should predominate. Equally if getting new women to try out your service is the issue then the initiation and engagement phases are critical. And if developing trial into retention and long-term loyalty are the primary problem then the focus should be on initiation, engagement and resolution. By ordering the identified tasks, marketers can begin to understand the logic flow of the service. We can also identify when each different colleague is involved and what their different roles are. Start with the colleague who has the first point of contact with the women and follow the line of activity until it reaches a change point. Ensure that each change point is framed as a yes/no type of action so that interactions among colleagues can be properly mapped. Include all colleagues in the service map who are involved.

2. Identify core customer expectations

The process map can now be used to identify which parts of the service are failing women and which parts are not effectively building strong

Task list		Customer expectations	Employee expectations
Task A. Booking a flight with a travel agent	What	Provide flight options	Provide advice and make arrangements
	How	Efficient and friendly	Efficiently and comprehensively
Task B.	What		
	How		
Task C.	What		
	How		
Task D.	What		
	How		

Figure 9.4 Mapping customer and employee expectations on key tasks across the customer journey.

relationships with them. These will help us to define the ideal service description. It is important to identify expectations from three different areas with both customers and colleagues. First we need to understand what a woman expects. Second we need to understand how a woman perceives the brand currently. Finally, we need to understand how a woman compares your brand with others she may have relationships with (see Figure 9.4).

We need to identify how well the current service is delivering against what women expect. Uncovering their expectations will provide terrific insights with which to build the ideal service. Expectations are different from needs in that they may be highly irrational and yet very important to women. For example, women who use the same airline route or hotel chain may expect that after a few trips the staff will remember their name and treat them differently. When using a phone banking service, women expect the phone to be answered within three rings or not to be put on hold. They may also expect that they will speak to a real person, rather than a machine, which is increasingly the case. As we identify expectations, we need to cluster them into different types; some will be about the "what" and some about the "how" of the service. They can be clustered into a series of buckets, for example:

- *Speed and efficiency* How quickly did the service engineer solve the problem or the car mechanic arrive at your roadside breakdown?
- *Style of service* How formal or engaging was the service compared with a women's expectations – was it overly familiar or too stuffy and formal?

- *Quality of the advice or knowledge* How accurate was the advice the travel agent gave about the location of the restaurant? How insightful was the advice from the pension advisor? How helpful was the advice from the lawyer in resolving your house sale?

By clustering the types of service expectations it is easy to determine the priority of activities within the relationship-building. Start with each task and identify the customer expectations for both the "what" and the "how." Then move on to the next task. Once this description is completed, we can identify work colleagues' expectations across the same range of activities. Comparing the two, it is easy to see where fundamental splits occur, leading to service lapses and relationship breakdowns. Women may reasonably expect a gasoline station attendant to know something about cars in case they break down, but the colleague's job description may not include the need for this type of experience. Or a woman may expect a restaurant waiter to know from which country the lamb in her meal comes or what ingredients are in the sauce on her dish. There is currently a move towards more localizing of service standards following decades of globalization of brands like Vodafone, IBM, Orange, HSBC, Starbucks, eBay and Amazon. While a global brand still needs to deliver the same perception to all its women customers, the very nature of service and customer relationship building – a people-delivered experience – means that it needs to be more locally relevant than globally standard. For example, a bank's female customers in India and Spain have very different attitudes towards money, saving for retirement and their respective economies' interest rates. Even the level of respect shown between the bank clerk and women may be different. The office opening hours may be different (Spanish companies often close for an hour or two in the afternoon and stay open later into the evening). In fact just about every dimension of human behavior may be perceived differently across different cultures. On the other hand, when women regularly experience the same brand across many different countries, it needs to be similar but not identical or the brand will be perceived as incoherent and lacking focus.

Defining a narrower target group of women creates stronger insights and propositions. Identifying and focusing on "queen bee" women will lead to a wide and profitable halo effect with other women.

When creating customer relationships with women it is important to take into account local attitudes towards service. These may rely on national stereotypes but it is easy to acknowledge that German service might be efficient but not warm or humorous, while Asian service is often warm, gracious and pleasing. Similarly, American service is of a high standard but relies too heavily on the overly obvious connection between good service and the size of tip you leave your waiter or cleaning lady. This may feel natural for Americans, but for Europeans it is often too gauche and suggests the people are only interested in the cash not the satisfaction of the delivery of fine service – although since one out of every eight Americans has worked in McDonald's at some point in their life, they certainly understand that when people pay for service they definitely expect it. Nationality-based service examples include the following:

- Southeast Asians are renowned for their ability to provide excellent service with a smile
- Singapore Airlines is often rated the best for pleasing and gracious service
- English service can be disappointingly poor for some nationalities; the staff are either superior and feel that service is beneath them or are over-polite to the point where the service is too intrusive
- Service in the Latin countries can be hit-or-miss to some nationalities – laid-back attitudes can come across as lack of interest in serving others, or at least not efficiently

A company's service ethos is the defining attitude towards serving the customer. While all companies like to think of themselves as customer-relationship-minded, very few have a truly feminine service ethos. First Direct bank has a down-to-earth and empathic service ethos. It prevents customers feeling patronized by treating them like normal human beings in an adult-to-adult conversation. It fosters fantastic relationships with its customers. A relationship-oriented service ethos like a travel agent creating individual vacations or retailers like Prêt à Manger will have greater appeal for women.

Feminine, relationship-oriented ethos

Pros

- Excellent for services that require a high degree of flexibility and need to be adapted constantly to changing requirements.

- Encourages colleagues to grow in themselves.
- Helps colleagues to feel valued through showing a high degree of trust in them.
- Encourages colleagues to focus on the individual customer.

Cons

- Requires more training and refreshers for colleagues.
- Needs higher degree of management trust.

Nordstrom, the US the fashion retailer, has dominated its market though the service it delivers, especially to women, its core customers. It has proved that women will pay a higher price for a superior service; in so doing it helps to differentiate them from low-price discount brands (Nordstrom's sales per square foot are twice the industry average).The service differentiation is delivered though the belief that at all times the most important person in the entire business is the customer; and this runs through all their channels (including online). The colleague manual is simple and contains only one rule: "Use your good judgement in all situations." (This is followed by "There are no additional rules.") Nordstrom recognize that building customer relationships or making friends is done one friend at a time. They share fantastic relationship-building stories throughout the organization to illustrate and train everyone about what their service ethos of going the extra mile means in practice. For example, a customer at their Chicago store was searching for a black bow tie:

> I was going to a black tie do, and needed a ready made bow tie. They didn't stock one, but the guy said – if you have ten minutes, how about I teach you how to tie one? And, in the middle of a heaving Saturday afternoon, he did just that and got the sale, I got happy, I recommend them to everyone and tell the story ten years later. Nordstrom have a few service components that keep their customers loyal.

- their staff come out from behind the till to hand over goods; they provide receipts in a wallet; and use a higher quality of bags
- detailed inventory and customer data to drive knowledge and service
- personal shoppers and on-site tailors
- spacious layouts and increased amenities (food/rest-rooms)

Creating customer moments of delight

Beyond the service ethos, which is a broad-based and often company-wide initiative, there are specific, high-impact icons that can demonstrate the brand and help cement relationships with women. These are typically focused on magic moments or a high-impact part of the customer journey. A truly magic moment is something that:

- changes the way the industry works
- creates powerful differentiation
- delivers against a deep customer insight
- improves customer satisfaction scores by double digits

Creating a magic moment is a key tool for differentiation for any brand. When Orange created "per second billing" it radically changed the landscape of mobile phones. Up until then, customers had always been charged by the minute. While this may seem small, those additional seconds actually cost customers a lot of money. UPS created a magic moment when they created their online package tracking function. This feature enabled the customer to easily track their own parcel live from their desktop, without having to check in constantly with the package helpline. Again, UPS created a magic moment changed the industry. We should not underestimate the value that a tangible icon or magic moment can bring in the definition of a new service regime. A physical totem accelerates reappraisal because it represents the new behavior or it acts as a way marker for staff and is a marketable icon towards female customers. Without such a tangible manifestation, is it sometimes difficult to convince women there has indeed been a change in service levels. Equally it is easy for staff to slip back into the bad old ways without a physical reminder of a new regime. Amazon, the online retailer, created a magic moment when they created "one-click shopping." Born out of an insight that regular users were unhappy about the length of the buying process, Amazon reversed the process and gained permission from customers to retain their pre-filled-in details so that they can purchase in one click when they return to the site. The service process maps for the original and the one-click shopping service are remarkably different and illustrate just how much time, effort and resources can be saved for both the customer and the vendor.

While it is important to improve the overall customer service through an ethos and the creation of magic moments, these need to be brand totems as well. The best way to assess how well these initiatives are at driving brand

Task list	Brand Positioning	Brand Value A.	Brand Value B.	Total	Comments
Booking ticket	2	3	3	3	Lacked personalization – didn't make me feel special
Check-in	5	4	4	4	Warm welcome, helped solve problem well
Lounge service					
Onboard					

Figure 9.5 Evaluating key customer experience components against fit with the brand positioning and values.

performance is to assess them against two criteria. First, they must have the ability to drive high awareness through catching a woman's eye. Second, they must be capable of accelerating the transfer of brand associations. These criteria can be assessed either qualitatively or quantitatively through research (see Figure 9.5).

Identify external benchmarks

Women customers will judge you against the best service they get from any provider, not just the ones in your category. In order to create a sustainable competitive edge you need to avoid incremental gains and focus on transformative improvements. The easiest way to achieve this is to look beyond your current category. Gaining insights from other countries is another great way to leapfrog the competition. Depending on the industry, it's easy to identify foreign brands that have transformed their service offer and enjoyed their female customers' loyalty, although this needs to be balanced with local understanding and cultural relevance – some ideas just don't travel well. Women take a more holistic point of view of the world than men, who are better able to compartmentalize their world. This is partly because women's brains are so much better connected and because they comprehend the interrelationships of actions much better than men, meaning that they will judge your brand and the kind of relationship and service delivery against all others in her world, not just those in the same category. Women will measure your performance against a range of service providers from every industry they encounter in their daily lives. So if Tesco or Wal-Mart has a brilliant helpline then they will use that as their benchmark for how their retail bank treats them, rather than just other poor service retail banks. It's no longer possible to be just the best in your category; you need to be the best in the world if you are to maintain your lead.

Marketing relationship scorecard

The relationship scorecard can be used to assess performance initially and to improve it continually in a dynamic market. The relationship scorecard should also be used to identify great benchmarks across the customer journey. This includes the pre- and post-journey beyond your actual service offer. Using the structure of the customer journey guarantees you will maximize the impact of any improvements. First, map out the customer journey, then using the scorecard identify benchmarks at each key point on the journey. The framework contains seven key areas or criteria to challenge your business. It is rare that a business will be exemplary in all these areas, but many will excel at several of them (see Table 9.1).

One way to identify the brands your target woman admires is to map out her brand service world. Identify her preferred brands for a range of services that are close both to your service and to her wider brand world. It is important to be as narrow as possible in your description of the female target audience, to ensure that results are actionable.

3. Define the ideal branded touchpoint experience

There are three objectives in creating the ideal service relationship.

- negate the dissatisfiers (close the gap between customer expectation and reality)
- operationalize the brand through service design
- enhance the satisfiers

Many firms focus on the last item, improving what they are good at already. However, often higher gains can be achieved with quicker impact by reducing the dissatisfiers and operationally embedding the brand in the service design. An initial, high-level cost–benefit analysis of all of these will prioritize activities. There are three inputs to this: brand, operational competencies, and customer needs. Businesses often focus on just one or two of these and never really maximize their potential. It is easy to see that if a mobile phone provider reduces its service waiting time by half then any woman will be happier. However, there is another layer to this improvement

Table 9.1   Marketing relationship effectiveness scorecard.

Criterion	Topic	Description
1. Who is the experience aimed at?	Audience	Target audience, segments, mindsets/mood states
2. What are their unmet needs?	Insights	Understanding true human attitudes and behaviors: avoiding stereotypes
3. Why does the brand exist in the world?	Positioning idea	Point of view on the world, relevant, differentiating, stretching, credible
4. How is the idea translated into service ethos/icons?	Ethos and icon concepts	Icons – powerful, tangible embodiment of the positioning idea
5. How do the icons create an experience?	Emotional qualities and icon design	Engaging, appealing, impact, significance for the customer, contemporary and simple
6. What is the marketing promise?	Setting expectations	Packaging and communicating the offer, messaging and design
7. What is the ideal customer outcome, actual outcome?	Audience takeout, ideal vs. actual	Emotional and aspirational feelings about the experience

that can create additional competitive advantage – the way this reduction is achieved can be on-brand or off-brand. For example, a medical service provider to US Hospitals needed to reduce their response time to improve hospital staff satisfaction. They established that the response time needed to be reduced by around 20 percent and identified a number of ways that they could achieve this:

- They could train the hospital staff to do more of the initial patient service themselves.
- They could hire more service personnel so they could reach more hospitals, more quickly.
- They could innovate the service process to reduce the length of time taken for each patient.

While all three of these would achieve the desired outcome, only one of them, the final option, would clearly demonstrate their brand value of "innovative medical expertise." In this case, the brand acted as a powerful decision-making tool and helped realign the relationship with the overall brand positioning.

The best way to activate the brand across the relationship is to cycle through the three components starting with customer needs, followed by competitive benchmarks, then the brand values. This should be undertaken for each step in the customer journey, including parts of the current journey that are prior and post to your brand delivering to the customer. This needs to be repeated several times to ensure maximum activation. It is more effective to undertake a prioritization after each round of brainstorming, so the list of service touchpoints is prioritized, in terms of impact ,into primary, secondary and tertiary. This also means that greatest effort is expended on the touchpoints that most affect the brand's relationship with a woman. Tesco the supermarket has a broad-based service ethos, "Every little helps." This pervades everything they do from improving the website to helping customers in store. There is no one big thing that differentiates them, but there are a collection of smaller things that when added together create a gestalt sense that the service is distinctively better. They were the first supermarket to open another checkout till if there were more than three people queuing at the current ones. They pioneered helpful service like escorting a shopper to their requested food item, rather than just telling them it was in aisle 14. They genuinely look happy to be at work and this kind of infectious vibe rubs off on customers.

> Focus on a couple of signature brand icons where it matters most rather than spreading the brand essence too thinly.

Ritz Carlton hotels use a colleague credo to reinforce their emphasis on building relationships. There is an umbrella message – "Ladies and gentlemen serving ladies and gentlemen" – that engenders a sense of pride in colleagues. There is also a service credo that contains 18 separate initiatives, which colleagues must learn and recite. To enhance the service ethos, there is a message from the CEO every day exalting colleagues to practice the credo: each day one specific item in the credo is chosen to be exemplified. This ensures that all 18 are continuously re-energized throughout an

18-day cycle. It also shows strong commitment from top management that long-term relationships really matter to the company's performance.

Lexus Automobile has succeeded in the upmarket segment of the US market that was dominated by European brands such as Mercedes and BMW. Although Japanese cars had been successful in the compact car market, none had been in the luxury sector until Lexus sold their first car in August 1989. The quality of their product goes far beyond the metal into unparalleled service though a carefully selected dealer network. Interbrand research in 2004 proved that this was still the case, with Lexus owners confirming a higher degree of satisfaction than from even Ferrari and Porsche. In the influential J D Powers customer satisfaction survey, Lexus regularly tops the list in the US and the UK. They use a variety of initiatives to ensure they make great friends with their customers:

- making waiting a pleasure in the dealership – from indoor driving range golf simulators (Oakhurst, New Jersey) to on-site manicurists (Las Vegas)
- valeting all cars on service
- pickup and delivery to home address the norm (not the exception)
- loan cars being always of comparable or better standard (with the option to try different models in the range)
- personal phone calls and letters for apologies
- on-line tools for customers to update their preferences and details

Ocado, the UK's upmarket online retailer, also pioneered online food shopping and delivery. They used their service ethos of "Every touch counts" to drive stronger relationships with women (most of their customers are female). Unlike other online shoppers that leave the shopping bags at the front door, they will carry the shopping bags to the kitchen table. They offer 30-minute delivery slots and phone up if they are going to be earlier or later. Should this occur, they carry a range of gifts, like a bottle of wine or new food item to give women as an apology. Here are other ways Ocado build friendships with their female customers:

- same route by drivers – so they get to know customers
- mercedes quality vans
- phone up if they are going to be late, even by 5 minutes
- bag of goodies for discretional use (for example, fruit and toys for children)

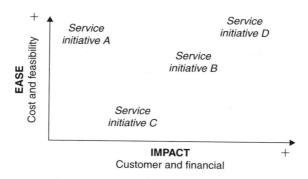

Figure 9.6 Interaction of ease and impact on outcome of investment in relationship-building.

- web site – send in your regular supermarket receipt and this will be programmed into the system so that regular purchase items can be highlighted

As we can see, certain areas of the customer journey are better for demonstrating certain brand values. It is better to tune the service delivery with specific values than try to incorporate all the values at each customer touchpoint. That way, they can shine in their own space, rather than having to compete with three or four other values or messages at the same time. A high-level cost–benefit analysis will ensure that a business invests money and resources on the things women will appreciate. There are often internal company myths about what is important to women and which things actually cost a lot of money and add a lot of value. An open mind means that we can challenge ourselves to review the relationship-building with objectivity. At this stage in the process a simple, high-level scorecard analysis is all that is required; later we will undertake a more detailed business case development. The primary criteria are ease and impact. Ease is a combination of low cost and good feasibility. High impact comprises impact on women's perceptions and behavior together with strong impacts on business results like margin, revenue and market share (see Figure 9.6).

4. Operationalizing the new touchpoint experience

Operationalizing the new touchpoint experience should include the creation of physical icons and compelling service experiences. Once the new relationship blueprint has been created, based on clearly defined outcomes, the relationship design needs to be mapped in detail. This involves the same mapping process used at the start of the virtuous cycle. Again, start at the beginning of a woman's interaction with your brand. Identify each step and decision point. Identify which colleague needs to interact with the customer at which point along the journey. Try to avoid adapting their current working practices and start with a clean sheet. Later these can be refined, but it is more important to first create the perfect service experience. At each touchpoint, define the content of the interaction – a brief description of the required tools/capabilities or conversation. This needs to describe both what the woman can expect and what is expected from the colleague.

5. Building the return on investment (ROI) business case

Any new initiative will require investment to realize any potential performance increase. There are many ways to define a business case or ROI for the new initiative. There are a few fundamentals that should always be included. The basic case has three components. First, the definition of the market size for the women you want to target. Second, the expected degree of improvement in those women's buying behavior. Finally, the cost and resources required of realizing those improvements for those women. These may be expressed in a number of ways but ideally as both percentages and absolute figures.

Sizing the target market

The market size in numbers of potential women customers needs to take into account their characteristics including any attitudinal boundaries. While it is often easy to get generic customer figures based on demographics and usage, it is much more difficult to get a clear view of the attitudinal segment. However, achieving this will enable you to be much more effective in your communications spend. This may increase the effectiveness of your above- and below-the-line spend by as much as 20 to 30 percent. It is therefore worthwhile investing some of that money in accurately researching the size of the market. It is quite difficult to assess the potential improvement in buying behavior by the target women. First, be very clear about what change is being sought and where this growth is likely to be achieved. Using the business growth matrix we can target specific types of business growth with different female customers and prospects. This can then be used to realign business and marketing activities to achieve higher than sector growth performance (see Figure 9.7).

221

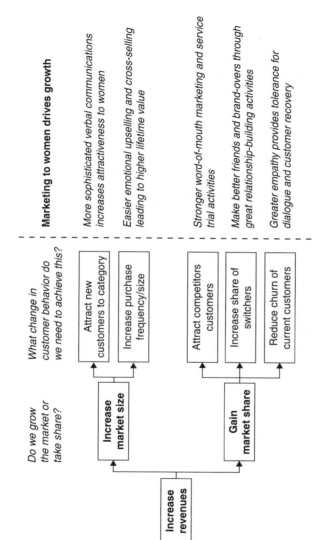

Figure 9.7 Examples of marketing to women initiatives linked directly to business growth strategies.

Clearly there are a number of ways to define the effect of the improvements. These range from light, internal estimates to a more rigorous statistical approach.

Light

A broad internal expert opinion may suffice. Typically a discussion with the senior sales team members will help identify possible gains. Usually these will be expressed as a growth percentage, for example 7 percent. Based on the market size and previous year's revenues/margins it is straightforward to convert these into financial forecasts.

Medium

The propensity to change behavior can be tested as part of qualitative and quantitative research. Presenting female customers with descriptions and visuals of the new service experience allows researchers to probe potential changes in favorability towards the new offer.

Heavy

Many of these effects can be modeled using sophisticated statistical techniques including conjoint analysis, demand driver analysis and touchpoint analysis. These require detailed data on female customer numbers and current buying patterns.

The expected change needs to be amortized and modeled over a number of years, typically between two and five depending on the business category. The expected uplift will not be linear, but any advantage will be diminished over the years as the competition improves.

Cost improvement analysis

There are two kinds of major cost: the capital expenditure cost (capex) involved and the ongoing cost of the changed initiative. A clear description of the changes will enable an accurate capex cost to be calculated. The cost of capital may need to be factored into this figure at the prevailing rate that the business uses. The ongoing cost will need to take into account the rising cost of labor over the period of costing. Typical capital expenditure (capex) costs include:

Internal
- creation of new relationship-building
- creation and implementation of a training program to educate staff

- development and production of service icons
- development and production of internal communications materials such as service ethos cards and so on
- legal and human resources (HR) costs for revising contracts and so on
- additional management capabilities

External
- development and production of new marketing materials
- research to confirm relationship-building and market potential

Typical ongoing costs will include:

- additional colleague costs
- additional tools to perform the new service for example call center scripts
- continuing training programs

There is a tendency for companies to see relationship-building as an additional cost. However, it can also be used to reduce costs and make the business more effective. It is rare that improving customer relationships saves large amounts of money in the short term, but long-term gains will be much higher. It is plausible that both these goals can be achieved with careful planning. Sensitivity to the absolute value of the improved customer relationship can be balanced against the break points in cost. It may be possible to achieve 80 percent of the improvement with less than 80 percent of the additional cost. It is therefore worth while to model the potential cost benefits in some detail in order to determine the most effective gain and resources required. The potential curve is probably a nonlinear relationship.

Improving staff relationships with women will almost always deliver the highest ROI and increase profitability.

Feasibility study

The business case also needs to identify the likely feasibility of the initiative. The feasibility should encompass all aspects of organizational change. These need to include:

- colleague skills and capabilities to deliver the new service
- management ability to embody and drive change
- capital availability at the appropriate time

- alignment and integration of new processes with old
- availability of tools to complete the job

Risk management of touchpoint improvements

It is important to identify all the relevant risks associated with the proposed changes. A standard risk management session will cover offer all the necessary aspects but these may include:

Internal
- lack of management buy-in.
- colleague change fatigue.
- health and safety issues.
- organization stress. (It is important to be wary about any organizational changes. Companies that attempt too much at once can find themselves paralyzed by the constant change or the magnitude of the changes required. The result can be chaos with everything up in the air. Generally the larger the organization, the more the impact of change will be compounded.)

External
- unexpected changes in the marketplace. (These may relate to sudden flooding of the market and its ensuing price crash for your service or product.)
- left field competition arriving in the marketplace. (New entrants may suddenly enter your market with a completely new business model that affects the whole market.)
- legislative. (Changes in the law or a new ruling can adversely affect the service offer.)

Pilot test each new touchpoint

Once the business case has been created and signed off, it is necessary to pilot test and implement the changes with the women the business hopes to attract. It is valuable to pilot test the new relationship design to ensure that it is truly building better relationships with these women. Pilot testing may have occurred prior to developing the business case or even in parallel with the business case. The pilot test may be more about a staged roll-out program to ensure that maximum learning is achieved with minimum negative impact on the organization. Working with the HR and training teams tests out the new service design on a small scale. Usually a few

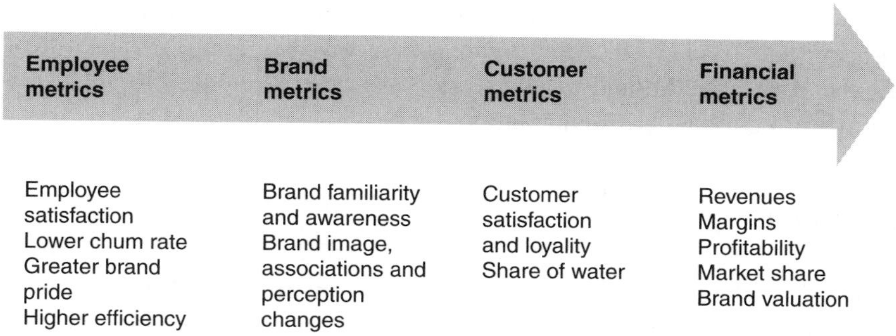

Employee metrics	Brand metrics	Customer metrics	Financial metrics
Employee satisfaction Lower chum rate Greater brand pride Higher efficiency	Brand familiarity and awareness Brand image, associations and perception changes	Customer satisfaction and loyality Share of water	Revenues Margins Profitability Market share Brand valuation

Figure 9.8 Key metrics employed to track touchpoint improvements.

weeks will be needed to both train and test. Frequently there will be further adaptations to the relationship design in the light of feedback from real women customers. Implementation plans need to be developed along with key operational ownership. It is advisable for all the key stakeholders to be involved from the start of the improvement process. But it is ultimately they that make a success of the changes, so the more that they can create the implementation plans the more likely it is that the plans will both work and be delivered.

6. Define performance metrics

There are four types of metrics that the marketing manager should implement to track touchpoint improvements. These are employee metrics, brand metrics, customer metrics and finally financial outcome metrics. As Figure 9.8 shows, each of these feed into each other to form a brand value logic chain. This is based on the service profit chain and connects the input of happy, motivated employees with satisfied customers who are willing to pay more money more often for their preferred brand. The financial metrics are crucial for increasing the credibility of marketing within the wider corporation and specifically with the executive board. The ability to translate softer marketing metrics like customer satisfaction into hard financial data is the only way to gain increasing support and investment in marketing.

Each of these needs to be tracked on a quarterly and annual basis if they are to provide effective input into the ongoing brand strategy. One or two of the key metrics should also form part of the overall business key performance indicators (KPIs) and certainly be a personal KPI for each board member. Without this level of integration into the general business process it is unlikely that the brand will receive the appropriate level of financial and operational investment and resources required to achieve a significant competitive advantage.

SUMMARY

Marketing channels need to be adapted to maximize connection with women audiences. These include greater use of word-of-mouth networks, more conversation websites and more-inclusive packaging designs.

Taking an analytical approach to high-impact touchpoint definition will help isolate prejudices and identify higher-impact ways to make stronger branded relationships with women.

Improving the ideal touchpoint experience for women requires a holistic approach from the entire company. Marketing should act as the primer to concept development and establish a robust business case, but this must be operationalized across every touchpoint if women are to believe it.

Bibliography

Aaker, David A. and Joachimsthaler, Erich (2000) *Brand Leadership: The Next Level of the Brand Revolution.* Free Press, New York.

AARP Public Policy Institute (2004) Research report, *The State of 50+ America.* aarp.org

Alder, Harry (1994) *NLP, The New Art and Science of Getting What You Want.* Piatkus, London.

Armstrong, Nicola (2006) *Trust Team Mum.* rethinkpink.com

Asian Development Bank (2004) "Key Indicators: Poverty in Asia: Measurement, Estimates and Prospects," cited in Chan Wee Guan and Marilyn Chew Su-Fern, *Women As Emerging Wealth Owners in Asia.* Esomar, London. auroravoice.com (2004)

Barletta, Martha (2003) *Marketing to Women.* Dearborn Trade Publishing, Chicago, IL.

Baron-Cohen, Simon (2004) *The Essential Difference.* Penguin, London.

Benson, Richard and Bilmes, Alex (2004) "Middle Class? Yes But Which Are You?" *Observer* (London), 3 October.

Betts, Kate (2005) *Boomer Chic: Why Are Speciality Retailers Like Gap and Chico's Targeting the Spending Power of Women of 35?* time.com

Blomqvist, Ralf; Dahl, Johan and Haeger, Tomas (2002) *Customer Relationship Development, Implementing Fast Track, First Base Customer Relationship Management Solutions.* Financial World Publishing, Canterbury.

Boomers International (2007) boomersint.org

Brand Keys (2007) *2007 Brand Keys Customer Loyalty Engagement Index.* brandkeys.com

Brandnoise (2006) *Baby Boomers and Brain Food.* brandnoise.com

Brizendine, Louann (2007) *The Female Brain.* Bantam, London.

Brody, L. R., and Hall, J. A. (2000) "Gender, Emotion, and Expression," in M. Lewis and J. M. Haviland (eds) *Handbook of Emotions.* Guilford Press, New York.

Brown, Mary and Orsborn, Carol (2006) *BOOM: Marketing to the Ultimate Power Consumer – The Baby Boomer Woman.* American Management Association, New York.

Bruce, D. and Bahrick, H. P. (1992) "Perceptions of Past Research," *American Psychologist* 47, 319–28.

Brunas-Wagstaff, Jo (1998) *Personality: A Cognitive Approach.* Routledge, London.

Buckley, Christine (2004) "Lady Luck Deserts Boardroom Women," *Times* 8 December.

Burns, Kate (MD of AdSales, Google UK) (2005) Speech at Rethink Pink Conference, London.

Burr, Gwynn (Director of Customer Service, Sainsbury's) (2005) Personal interview.

Burr, Vivien (1998) *Gender and Social Psychology*. Routledge, London.

BusinessWeek (2004) *Top Tips for Marketing to Women*. businessweek.com

BusinessWeek (2005) "I am Woman, Hear Me Shop," special report ed. Patricia O'Connell, February 14.

BusinessWeek and Gallup (2004) Cited in Joanna L. Krotz (2006) *Women Power: How to Market to 51% of Americans*. microsoft.com/small business

BusinessWeek/Interbrand (2006) *Best Global Brands 2006 League Table*. BusinessWeek/Interbrand, New York.

BusinessWeek/Interbrand (2007) *Best Global Brands 2007 League Table*. BusinessWeek/Interbrand, New York.

Byrne, D. (1971) *The Attraction Paradigm*. Academic Press, New York.

Byrne, D., Clore, G. and Smeaton, G. (1986) "The Attraction Hypothesis: Do Similar Attitudes Affect Anything?" *Journal of Personality and Social Psychology* 51. 1167–70

Cairns, Warwick (Planning Director, Brandhouse WTS) (2006) Personal interview.

Calne, Donald, B. (Professor of Neurobiology, University of British Colombia) (2005) In Saatchi & Saatchi presentation, Rethink Pink Conference, London.

Canli, Turhan (2002) "Women's Better Emotional Recall Explained," *New Scientist* 22 July.

Carat (2005) *Project Britain: Segmenting Singletons*. Carat Media Agency research results. singledom.co.uk

Carat (2006) *Born to be Wired*. http://www.carat.com/carat/IntranetDocViewer? wsDocTypeId=0&wsScreenType=95&wsRow=1&wsCol=6&wsDepth=1& wsBI=null

Carter, Rita (1998) *Mapping the Mind*. University of California Press, London.

Caterall, Miriam and Maclaren, Pauline (2002) "Gender Perspectives in Consumer Behavior: An Overview and Future Directions," *Marketing Review* 2, 405–25.

CCMC (Communications Consortium Media Center) (2007) *Equality 2020*. ccmc.org

chinadailynews.com (2006) *Most Chinese Mobile Users Sure to Buy 3G Handsets*.

chinamobile.com (2007)

cia.gov (2007) *The World Fact Book 2007*.

Clegg, B. (2000) *Capturing Customers' Hearts: Leave the Competition to Chase Their Pockets*. Financial Times/Prentice Hall, London.

Cody, M.J. and McLaughlin, M.L. (eds) (1990) *The Psychology of Tactical Communication*. Multilingual Matters, Clevedon.

Collins, Jim (2001) *Good to Great.* Random House, London.

Connell, R.W. (2002) *Gender.* Polity Press, Cambridge.

Coren, Stanley, Ward, Lawrence M. and Enns, James T. (1994) *Sensation and Perception.* Harcourt Brace, Fort Worth, TX.

Csikszentmihalyi, Mihaly (1990) *Flow: The Psychology of Optimal Experience.* Harper & Row, New York.

Danis, Fran S. (2005) *Flora's Vest: The Personal Is Political Is Personal Again.* Sage, London.

Denny, Linda (2006) Quoted in Joanna L. Krotz, *Women Power: How to Market to 51% of Americans.* microsoft.com/small business

Dove (2006) *Beyond Stereotypes; Rebuilding the Foundation of Beauty Beliefs – Dove Global Survey.* campaignforrealbeauty.co.uk

Dwyer, Diana (2000) *Interpersonal Relationships.* Routledge, London.

Eagly, A.H. (1987) *Sex Differences in Social Behavior: A Social Role Interpretation.* Erlbaum, Hillsdale, NJ.

Eagly, A.H. and Warren, R. (1976) "Intelligence, Comprehension and Opinion Change," *Journal of Personality* 44, 226–42.

Eagly, A.H., Wood, W. and Fishbaugh, L. (1981) "Sex Differences in Conformity: Surveillance by the Group as a Determinant of Male Conformity," *Journal of Personality and Social Psychology* 40, 384–94.

Eatson, Neil (of Zed, New Media) (2006) Cited in Sean Hargrave, "Men Flick, Women Stick When Online," *Marketing Week* 3 February.

Economist (2006) "The Importance of Sex. Forget China, India and the Internet: Economic Growth Is Driven by Women," 12 April.

Egan, John (2001) *Relationship Marketing.* Pearson Education, London.

Ellwood, Iain (2001a) *Seeing Purple.* Economist, London.

Ellwood, Iain (2001b) *Fine Teas and Poor Management.* Economist, London.

Ellwood, Iain (2001c) *The Leisure Principle.* Financial Times, London.

Ellwood, Iain (2002) *The Essential Brand Book.* Kogan Page, London.

Ellwood, Iain (2006a) *Global–Local Banking.* Bank Marketing International, London.

Ellwood, Iain (2006b) *Promoting the Paralympic Brand.* Marketing Week, London.

emarketer (2007) *More Women Online. Women Outnumber Men Online and It's Likely to Stay That Way.* April 9 2007. emarketer.com

EOC (Equal Opportunities Commission) (1996) *Facts About Women and Men in Great Britain.* Manchester.

Factiva (2001) *Media Usage Reflects Emotional Language Increase in Great Britain.* factiva.com

Feirling, Alice (2006) *Of Wine and Women – Serious Wine Collecting Is No Longer a Male Sport. Why Don't Marketers Get It?* ftimes.com

Fisher, Robert J. and Dube, Laurette (2005) "Gender Differences in Responses to Emotional Advertising: A Social Desirability Perspective," *Journal of Consumer Research* 31 (March), 850–8.

forbes.com (2006) *The 100 Most Powerful Women.*

Fragiacomo, Laura (2006) *Talking About Y Generation.* misweb.com

Francese, Peter (2003) "Ahead of the Next Wave – Generation Y," *American Demographics* 1 September.

Franzen, Giep and Bouwman, Margot (2001) *The Mental World of Brands.* World Advertising Research Centre, Henley-on-Thames.

Fubon Bank (2005) *Fubon Bank and Bonluxe Jointly Launch the Bonluxe Visa Which Turns Every Female Cardholder into Miss Perfect.* fubonbank.com.hk

Furnham, Adrian (1986) *All in the Mind: The Essence of Psychology.* Whurr, London.

Furnham, Adrian and Heaven, Patrick (1999) *Personality and Social Behaviour.* Arnold, London.

Gahilan, A.T. (2003) "The Amazing Jetblue," *Fortune Small Business* 13(4) 51–3

Goldman Sachs (2007) "Women-Led Stocks Have Outperformed the Market," in Kevin Daly, *Gender Inequality, Growth and Global Ageing*, Global Economics Paper 154. ftd.de/wirtschaftswunder/resserver.php?blogId=10&resource=globalpaper154.pdf

Goleman, Daniel (1996) *Emotional Intelligence: Why It Can Matter More Than IQ.* Bloomsbury, London.

Gordon, Kim T. (2002) "Chick Magnet; Tactics: How Can You Attract Women to Your Business?" *Entrepreneur*, March.

Gordon, Wendy (2004) "Think Pink" Conference, London.

Gray, John (1992) *Men Are From Mars, Women Are From Venus: The Definitive Guide to Relationships.* Element, London.

Gregory, Richard L. (1998) *Eye and Brain: The Psychology of Seeing.* Oxford University Press.

haier.com (2007)

Hargrave, Sean (2006) "Men Flick, Women Stick When Online" – *Marketing Week* 3 February.

Heathfield, Susan M. (2008) *Listen with Your Eyes. Tips for Nonverbal Communication.* About.com:Human Resources. Accessed 20 August.

Herrmann, D. and Crawford, M. (1992) "Gender-Linked Differences in Everyday Memory Performance," *British Journal of Psychology* 83, 221–31.

Hines, JoAnn (2006) *Boomers Are A Booming Business – Why Boomers?* http://www.packagingdiva.com/articles.htm

Hofsteede, Geert (1994) *Cultures and Organisations: Intercultural Cooperation and Its Importance for Survival.* HarperCollins, London.

honda.com (2006) *Honda Sees Global Sales Hitting Records.*

IMRB (2007) *Indian Consumer Products Research Focus Groups*. Mumbai.

Interbrand (2007) "Role of Brand Analysis," Graham Hales (Global Communications Officer). Personal interview.

internetworldstats.com (2007)

Jana, Reena (2005) "Will Women Jump at Exergames?" *BusinessWeek* 25 November.

Johnson, Lisa and Learned, Andrea (2004) *Don't Think Pink*. American Management Association, New York.

Kirk, Charles, Sgries, Jorg and Laursen, Mogens (2006) *Business Opportunity Number 1: Women in the Automotive Industry*, Esomar World Research Paper. esomar.org

Knoploch, Zilda, Wallis, Jem and Marjenberg, Rob (2005) *Learning About Consumers Through a New Bricolage*, Esomar World Research Paper. esomar.org

Korn, Melissa, (2005) "Women and the Art of Product Development," *Fast Company* 19 October.

Kovecses, Zoltan (2003) *Metaphor and Emotion: Language, Culture and Body in Human Feeling*. Cambridge University Press.

Krotz, Joanna L. (2006) *Women Power: How to Market to 51% of Americans*. microsoft.com small business

Laney, Emma (2005) "The Divine Secrets of Marketing to the Sisterhood," Speech at ReThink Pink Conference, London.

Lawson, Gillem and Brahma, Sunanda (2006) *Women's Views on Their Portrayal in Advertising. We've Changed. Do Advertisers Know?*, Esomar World Research Paper. esomar.org

Learned, Andrea (2002) *The Six Costliest Mistakes You Can Make in Marketing to Women*. marketingprofs.com

Learned, Andrea (2005) *Love Sweet Love: DHL Resonates with Women's Market*. learnedonwomen.com

LeDour, Joseph (Professor of Neuroscience, New York University) (2005) In Saatchi & Saatchi presentation, ReThink Pink Conference, London.

Lee, Louise (2005) "Beautiful Boomers Are Rewinding Not Winding Down," *BusinessWeek* 24 October.

Lee, Margaret T. and Ofshe, Richard (1981) "The Impact of Behavioral Style and Status Characteristics on Social Influence: A Test of Two Competing Theories," *Social Psychology Quarterly* 44(2) 73–82.

Leeds-Hurwitz, Wendy (1993) *Semiotics and Communication: Signs: Codes, Cultures*. Erlbaum, London.

Legato, Marianne, J. (2005) *Why Men Never Remember and Women Never Forget*. Rodale, London.

Leibenluft, E. (1998) "Why Are So Many Women Depressed?" *Scientific American* June.

Lenz, Elinor and Myerhoff, Barbara (1985) *The Feminization of America: How Women's Values Are Changing Our Public and Private Lives*. St Martin's Press, New York.

Levinger, G. (1980) "Toward the Analysis of Close Relationships," *Journal of Experimental Social Psychology* 16, 510–44.

Lidwell, W., Holden, K. and Butler, J. (2003) *Universal Principles of Design*. Rockport Publishers, Gloucester, MA.

Lindstrom, Martin with Seybold, Patricia B. (2003) *Brand Child*. Kogan Page, London.

Lippa, Richard A. (1994) *Introduction to Social Psychology*. Brooks-Cole, Pacific Grove, CA.

Maclaren, Pauline and Catterall, Miriam (2000) "Bridging the Knowledge Divide: Issues on the Feminization of Marketing Practice," *Journal of Marketing Management* 16, 635–46.

MacroMonitor (1999) "Women as Financial Consumers: Gaining Ground," *MacroMonitor Marketing Report*, January, vol. IV, no. 2. http://www.sric-bi.com/CFD/MRsummaries/MR.IV-2.shtml

Maister, David, H. (2001) *Practice What You Preach: What Managers Must Do to Create a High Achievement Culture*. Free Press, London.

Maslow, Abraham H. (1987) *Motivation and Personality*. Longman, Reading.

Metlife (2005) *The MetLife Market Survey of Nursing Care and Home Care Costs September 2005*. Metropolitan Life Insurance Company, New York.

Meyers-Levy, Joan (1988) "The Influence of Sex Roles on Judgment," *Journal of Consumer Research* 14, 522–30.

Meyers-Levy, Joan (1991) "Exploring Differences in Males' and Females' Processing Strategies," *Journal of Consumer Research* 18, 63–70,

Milligan, Andy and Smith, Shaun (2002) *Uncommon Practice: People Who Deliver a Great Brand Experience*. Interbrand–Pearson Education, London.

MLC (2007) Marketing Leadership Council website: mlc.executiveboard.com.

Moir, Anne and Jessel, David (1989) *Brain Sex: The Real Difference Between Men and Women*. Delta Publishing. Surrey.

Moss, Gloria (2003) "The Implications of the Male and Female Design Aesthetic for Public Services," *Innovation Journal* 8(4) http://www.innovation.cc/discussion-papers/moss-gender.pdf

Moss, Gloria and Colman, Andrew (2001) "Choices and Preferences: Experiments in Gender Differences," *Journal of Brand Management* 9(2), 89–98.

Myers, Gerry (2006) *The Bottom Line Case for Marketing to Women*, parts 1 and 2. advisorylink-dfw.com

NAS (2006) *Generation Y, Ready or Not, Here They Come*. nas.com

New York Times (2001) "Work First, Invest Later? Not These Days, to Be Old, Gifted and Employed Is No Longer Rare," *New York Times*, 14 January.

Newspaper Association of America (NAA) (2004) *Retail Rituals: Women's Changing Attitudes Toward Shopping.* naa.org

NFL (2005) Susan Rothman (Vice President for Consumer Products), *Sports Product Sales,"* nfl.com

Nielsen (2005) *Net Ratings, Home and Work Data, UK, June.* London.

nVision/Future Foundation (2003) *Changing Lives: Media and Gender Survey, UK.* futurefoundation.net

Onya-san (Marketing Director, ANA Hotels Group) (2006) Personal interview.

Oppenheimer Funds (2006) *What Every Baby Boomer Needs to Know About Planning for Retirement. Women and Investing.* oppenheimerfunds.com

Otaki, Midori *et al.* (1986) "Maternal and Infant Behaviour in Japan and America," *Journal of Cross-Cultural Psychology* 17(3), 251–68.

Paul, Vivek (2005) *India Going Global: India's Rapidly Growing Influence in International Markets.* asiasociety.org

Pease, Allan and Barbara (2001) *Why Men Don't Listen and Women Can't Read Maps.* Orion, London.

Pease, Allan and Barbara (2004) *The Definitive Body Language Guide.* Orion, London.

Peek, Laura (2004) "Online Casinos Lure Women into Hidden Addiction," *Times* 11 September.

Pervin, Lawrence, A. and John, Oliver, P. (2001) *Personality: Theory and Research.* Wiley, Chichester.

Peters, Tom (2005) *Women Roar! Re-imagine.* DK Adult, London.

Pisarkiewicz, Mary F. (2006) *Women, Cars and Branding.* designpm.com

Popcorn, Faith and Marigold, Lys (2001) *Eveolution: The Eight Truths of Marketing to Women.* HarperCollins, London.

Project Britain (2005) projectbritain.com

prudential.com.sg (2007) Prusmart Lady life insurance.

Putrevu, Sanjay (2001) "Exploring the Origins and Information Processing Differences Between Men and Women: Implications for Advertisers," *Academy of Marketing Science Review* 20, 1–14.

Ragone, Elizabeth (2000) "Marketing to Women on the Web," conference by the Marketing Institute, New York, 24–25 February.

Rank, Hugh (1976) "Teaching about Public Persuasion," in Daniel Dietrich (ed.) *Teaching about Doublespeak.* National Council of Teachers of English, Urbana, IL.

ReThink Pink Conference (2005) Introduction to Conference, London.

Robarts, Guy (2005) *"Gender Blur" Impacts Sales Tactics.* bbc.co.uk

Rodenburg, John (2003) *What Do Women Really Want Online? The Role of Magazine Brands on the Internet,* Esomar World Research Paper. esomar.org

Rogers, Lois (2005) "Boys, You Can Keep Your Greasy Pole," *Sunday Times* 12 June.

Rosen, Emanuel (2001) *The Anatomy of Buzz: Creating Word of Mouth Marketing*. HarperCollins, London.

saga.co.uk (2006) *Saga Magazine and the Over 50's*.

Salmon, Clare (Director of Marketing and Commercial Strategy, ITV) (2005) "Something for the Ladies," speech at ReThink Pink Conference, London.

Santee, R.T. and Jackson, S.E. (1982) "Sex Differences in Evaluative Implications of Conformity and Dissent," *Social Psychology Quarterly* 45, 121–5.

Scase, Richard (2000) *Britain in 2010: The New Business Landscape*. Capstone, London.

Schacter, D.L. and Scarry, E. (eds) (2000) *Memory, Brain and Beliefs*. Harvard University Press, Cambridge, MA.

Schaffner, Dionn (2000) *Marketing to Women on the Web*. garden.com

Schwartz Peter (2006) "How to Ride the Hottest New Trends," *Business 2.0*. cnn.com

Siennicki, Judy (2000) *Gender Differences in Nonverbal Communications*. colostate.edu

Skene-Johnson, Olive (2003) *The Sexual Rainbow*. Fusion, London.

Stallwood, Oliver (2005) "Why Women Find Parking So Tricky," *Metro*, January 24.

Tannen, D. (1990) *You Just Don't Understand: Women and Men in Conversation*. Ballantine, New York.

Thwaites, Tony, Davis, Lloyd and Mules, Warwick (1994) *Tools for Cultural Studies: An Introduction*. Macmillan Education, Melbourne.

Time (2000) "Early Puberty: A Growing Epidemic," 24 October.

Time Asia (2000) "The Ranks of Revolutionaries," 23 October.

timeasia.com (2006) *The Next Generation: China*.

titlenine.com (2007) *Our History: US Sports Legislation*.

Tuthill, D. and Forsyth, D.R. (1982) "Sex Differences in Opinion Conformity and Dissen," *Journal of Social Psychology* 116, 205–10.

Tweedale, Sophie (2004) "What Women Really Want," *Sunday Times*, 29 February.

US Census Bureau (2001) Report 952 *Highlights of Women's Earnings in 2000*. US Department of Labor, Washington, DC.

US Census Bureau (2003) *Generation X Speaks Out on Civic Engagement and the Decennial Census*. census.gov

US Census Bureau (2005) *Educational levels*. Income, Earning and Poverty data from the 2005 community survey. US Department of Commerce, Washington, DC.

Van Someren, Megan (2006) Quoted by Marshall Lager in "X Ways," *CRM Magazine* November.

waitrose.com (2006) Accessed 9 August.

Wall Street Journal (2006) "DIY Spending by Women Outpaces Men," 26 September.

Watson, D. and Tellegen, A. (1985) "Toward a Consensual Structure of Mood," *Psychological Bulletin* 98, 219–35.

Wee Guan, Chan and Chew Su-Fern, Marilyne (2005) *Women As Emerging Wealth Owners in Asia*. Esomar, London.

Wellner, Alison Stein (2003) "The New Adults: Gen Xers Will Turn 38 This Year," *Forecast* January.

Westen, Drew (2002) *Psychology, Brain, Behavior and Culture*. Wiley, New York.

Widhalm, Shelley (2006) *Generation Attitude Gap*. washingtontimes.com

Williams, J.E. and Best, D.L. (1982) *Measuring Sex Stereotypes: A Thirty-Nation Study*. Sage, Beverly Hills, CA.

Wong, Edward (2001) "Nike Enters the Gender Debate," *New York Times* 19 June.

Wreden, Nick (2005) *Women and Branding: Time for a Change*. brandchannel.com

Yates, Andy (2005) *Women Are Better Investors than Men*. digitallook.com

Yin, Sara (2006) "Motorola Bids for New 'Lifestyle' Sell," brandrepublic.com

Zaltman, Gerald (2003) *How Customers Think: Essential Insights into the Mind of the Market*. Harvard Business School Press, Boston, MA.

Zhang, Jing and Shavitt, Sharon (2003) "Cultural Values in Advertisements to the Chinese X-Generation," *Journal of Advertising* 32(1), 23–33.

Index